ALMA MATER STUDIORUM – UNIVERSITY OF BOLOGNA
DEPARTMENT OF ELECTRICAL, ELECTRONIC, AND INFORMATION
ENGINEERING "GUGLIELMO MARCONI" (DEI)

MATRIX DESIGNS AND METHODS FOR SECURE AND EFFICIENT COMPRESSED SENSING

A doctoral dissertation by
VALERIO CAMBARERI

SUPERVISORS:
Prof. RICCARDO ROVATTI
Prof. GIANLUCA SETTI

COORDINATOR:
Prof. ALESSANDRO VANELLI-CORALLI

PH.D. PROGRAMME IN ELECTRONICS, TELECOMMUNICATIONS AND INFORMATION TECHNOLOGIES ENGINEERING
XXVII CYCLE, JANUARY 2012 – DECEMBER 2014
S.C. 09/E3, S.S.D. ING-INF/01

This doctoral dissertation is submitted in partial fulfilment of the requirements for the degree of Doctor of Philosophy in Electronics, Telecommunications and Information Technologies, University of Bologna, Italy.

Dissertation Reading Committee:
Prof. R. Rovatti (Supervisor, University of Bologna, Italy)
Prof. G. Setti (Co-Supervisor, University of Ferrara, Italy)
Prof. L. Jacques (Reader, Université Catholique de Louvain, Belgium)
Prof. J.M.F. Moura (Reader, Carnegie Mellon University, U.S.A.)
Prof. S. Theodoridis (Reader, National and Kapodistrian University of Athens, Greece)

This dissertation will be defended on May 19th, 2015 at the University of Bologna – Faculty of Engineering, Viale Risorgimento 2, Bologna.

ISBN: 978-1-326-22687-9
All unpublished material in this document is © 2015 of the author, all rights reserved. All published material is © 2013-2015, IEEE.

*"Picard, you are about to move
into areas of the galaxy filled with
wonders you cannot possibly imagine...
And terrors to freeze your soul!"*

– Q to Captain Jean-Luc Picard, in "Q Who?",
Star Trek: The Next Generation, 1989.

A te che hai la chiave,
EQIM ICRP ILKO CLSF
ZJXK MGFV PQVX JCIC
XMZW HWKG ZFTI PMR.

Abstract

THE idea of balancing the resources spent in the acquisition and encoding of natural signals strictly to their intrinsic information content has interested nearly a decade of research under the name of compressed sensing. In this doctoral dissertation we develop some extensions and improvements upon this technique's foundations, by modifying the random sensing matrices on which the signals of interest are projected to achieve different objectives.

Firstly, we propose two methods for the adaptation of sensing matrix ensembles to the second-order moments of natural signals. These techniques leverage the maximisation of different proxies for the quantity of information acquired by compressed sensing, and are efficiently applied in the encoding of natural signals with minimum-complexity digital hardware.

Secondly, we focus on the possibility of using compressed sensing as a method to provide a partial, yet cryptanalysis-resistant form of encryption. In this context, we also show how a random matrix generation strategy with a controlled amount of perturbations can be used to distinguish between multiple user classes with different quality of access to the encrypted information content.

Finally, we explore the application of compressed sensing in the design of a multispectral imager by implementing an optical scheme for compressive imaging. This design entails a coded aperture array and Fabry-Pérot spectral filters. The signal recoveries obtained by processing real-world measurements show promising results, that leave room for an improvement in terms of an accurate calibration of the sensing matrix as applied by the devised imager at the optical level.

Acknowledgements

APPROACHING the defence of a doctoral thesis is a privilege and a magical moment: after slightly more than three years of rushing between paper drafts, absurd reviews, small successes and plenty of unfinished ideas, it's time to look back and thank the people who have led me in this here and now.

First and foremost, this thesis would not have been possible without the *guides*: Prof. Riccardo Rovatti and Prof. Gianluca Setti. You have supervised me in my efforts with attention and sheer interest, and have been two very distinct, yet equally great role models, that in many senses have shaped and enriched my scientific character.

In this adventure of a thesis, I also had many great *comrades* who made laboratory life challenging, stimulating and fun: Salvatore Caporale, Mauro Mangia, Fabio Pareschi, Prof. Sergio Callegari, Carlos Formigli and Javier Haboba. Dear friends, I did learn from you a lot; sharing doubts, trivia, scepticism, different attitudes and brilliant ideas was an important part of this experience that, I hope, has left you nearly as much as it has left me.

During my 2014 visiting period at IMEC in Leuven, Belgium (the first research experience abroad, leaving me wanting for more) I have been hosted by incredibly kind, friendly and expert scientists. It really was a defining experience, for which I would like to thank Andy Lambrechts, Bert Geelen, and the rest of the Integrated Imaging team (Klaas, Pilar, Bart, Carolina, Murali, Geert, Richard). At that time, I also had the pleasure of working with Prof. Gauthier Lafruit, who involved me to participate in the fascinating endeavour that is only partly reported in the final Chapter of this thesis; the same gratitude goes to Prof. Laurent Jacques (who also kindly accepted to review this thesis) and Kévin Degraux.

I would also like to thank the Reading Committee for providing many helpful comments, as well as suggestions for future improvements upon this thesis.

On a more personal note, my whole family (a certain black cat included) has been supportive throughout this experience, and deserves enormous thanks. I finally apologise to my dearest friends and loved ones, for showing understanding when I devoted more and more spare time to finishing papers rather than nurturing our relationships and shared passions; thank you for making every bit of time unique.

Table of Contents

List of Figures   xvi

List of Statements   xix

List of Acronyms   xxiii

List of Symbols   xxvi

Introduction   1

1   Elements of Sparse Signal Models and Compressed Sensing   7
     1.1   Low-Dimensional Signal Models   11
     1.2   Sensing Operators and their Properties   23
     1.3   Signal Recovery Algorithms   36

I   Adaptive and Efficient Matrix Designs for Compressed Sensing   51

2   Maximum Energy Sensing Matrix Designs for Localised Signals   53
     2.1   Sparse and Localised Signals   54
     2.2   Rakeness and the Rationale of Energy Maximisation .   58
     2.3   Maximum Energy Sensing Matrix Designs   62
     2.4   Performance Evaluation   73

3   Low-Complexity Digital Signal Compression by Compressed Sensing   77
     3.1   Lossy Compression Schemes for Biosignals   79
     3.2   Performance Evaluation   86

4 Maximum Entropy Sensing Matrix Designs for Localised Signals — 93
- 4.1 Compressed Sensing with Deterministic Ensembles — 94
- 4.2 Entropy Considerations on Localised Signals — 95
- 4.3 Maximum Entropy Sensing Matrix Designs — 96
- 4.4 Performance Evaluation — 99

II Low-Complexity Security by Compressed Sensing — 107

5 Average Recovery Performances in the Presence of Random Matrix Perturbations — 109
- 5.1 Compressed Sensing with Matrix Perturbations — 110
- 5.2 Average Performances with Matrix Perturbations — 113
- 5.3 Performance Evaluation — 115

6 Low-Complexity Multiclass Encryption by Compressed Sensing — 123
- 6.1 Principles of Multiclass Encryption — 125
- 6.2 Recovery Error Guarantees and Bounds — 131
- 6.3 Performance Evaluation — 142

7 A Statistical Cryptanalysis of Compressed Sensing — 153
- 7.1 Security Limits — 153
- 7.2 Achievable Security Properties — 155

8 A Computational Cryptanalysis of Compressed Sensing — 167
- 8.1 A Theory for Known-Plaintext Attacks — 168
- 8.2 Signal Recovery-Based Class-Upgrade Attacks — 188
- 8.3 Performance Evaluation — 192

III A Multispectral Compressive Imager by Random Convolution — 199

9 A Multispectral Compressive Imager by Random Convolution — 201
- 9.1 Imaging by Random Convolution — 203
- 9.2 Out-of-Focus Random Convolution — 208
- 9.3 Design and Implementation — 215
- 9.4 Performance Evaluation — 220

Conclusions — 229

Bibliography 235

List of Figures

1.1 A standard Nyquist-rate sensing scheme   9
1.2 A generic compressed sensing scheme   9
1.3 The mathematical elements that define the framework of compressed sensing 10
1.4 The "Fresh Fruit" MS sample image 19
1.5 Representation of a MS image on a sparsity basis obtained by Kronecker product of a 2D Haar DWT and a DCT basis 20
1.6 Evaluation of the compressible and low-rank behaviour of a MS image 21
1.7 Empirical distribution of the singular values of the RGE . . 31
1.8 Empirical distribution of the singular values of the RBE . . 32
1.9 The Donoho-Tanner phase transition curve 35
1.10 Geometric interpretation of LLS and OLS 43
1.11 Geometric interpretation of BP and BPDN 44

2.1 A centred multivariate Gaussian mixture defined on two subspaces 56
2.2 Probability of successful recovery by a selection of maximum energy measurements 61
2.3 Synthesis of maximum energy random sensing matrices . 67
2.4 Empirical distribution of the singular values of an exemplary aRGE 69
2.5 A comparison of the sequences of eigenvalues output from the solution of maximum energy optimisation problems . 70
2.6 The minimum number of measurements needed by an exemplary maximum energy sensing matrix design to attain probability 0.9 of successful reconstruction 72
2.7 Empirical phase transition boundaries for maximum energy aRGE *versus* RGE 74

List of Figures

3.1 A standard sensor node – processing node pair 78
3.2 Block diagram of the three evaluated compression schemes 80
3.3 A digital hardware implementation of the CS encoding stage with RBE matrices 83
3.4 Correlation matrices related to a maximum energy aRBE encoding matrix design 85
3.5 Average SNR for a lossy compression based on CS with different encoding matrices and quantisation policies . . . 88
3.6 Achieved code rates of the evaluated compression schemes: first quantisation policy 90
3.7 Achieved code rates of the evaluated compression schemes: second quantisation policy 91

4.1 Empirical PDFs of two orthonormal projections of natural images and their Gaussianity 101
4.2 Signal recovery of handwritten digits from maximum and non-maximum entropy sensing matrices 102
4.3 Signal recovery of electrocardiographic tracks from maximum and non-maximum entropy sensing matrices 104

5.1 MSNR curves used to set m beyond the phase transition of a perfectly informed BP 116
5.2 Comparison of the MSNR estimate with the recovery performances under DGA perturbation 118
5.3 Comparison of the MSNR estimate with the recovery performances under DUM perturbation 119
5.4 Comparison of the MSNR estimate with the recovery performances under SSF perturbation 120

6.1 An overview of two-class encryption by CS 126
6.2 Encoding matrix generator architecture 130
6.3 A single-transmitter, multiple-receiver multiclass CS network 130
6.4 Empirical evaluation of the perturbation constants for a two-class recovery error upper bound 140
6.5 Multiclass CS of speech signals 145
6.6 Multiclass CS of ECG signals 147
6.7 Multiclass CS of images 150

7.1 Outcome of second-level KS statistical tests to distinguish between two orthogonal plaintexts x', x'' 159

7.2 Estimated Kullback-Leibler divergence between the probability distributions of two ciphertext elements corresponding to different original signals 161
7.3 Empirical evaluation of the convergence rate constant . . 163

8.1 A two-class encryption scheme and the known-plaintext attacks being analysed from an eavesdropper (Eve) and a second-class user (Steve) 170
8.2 Empirical average number of solutions for Eve's KPA compared to the theoretical approximation of (8.4) 174
8.3 Gaussian approximation of the expected number of solutions 176
8.4 Empirical average number of solutions for Eve's KPA at Hamming distance h from the true one, compared to the theoretical approximation of (8.9) 182
8.5 Empirical average number of solutions for Steve's KPA compared to the theoretical approximation of (8.16) . . . 186
8.6 Average SNR performances of a class-upgrade known-ciphertext attack using signal recovery under matrix uncertainty algorithms 190
8.7 Effectiveness of Eve and Steve's KPA in recovering a hidden ECG 194
8.8 Effectiveness of Eve and Steve's KPA in recovering hidden image blocks 196

9.1 A FP-filtered sensor array 203
9.2 Optical scheme of a panchromatic compressive imager by out-of-focus random convolution 208
9.3 Optical scheme of a multispectral compressive imager by out-of-focus random convolution 210
9.4 Simulated recovery performances of a multispectral compressive imager by out-of-focus random convolution . . . 213
9.5 Pictures of the designed imaging system on the optical table 217
9.6 Point Spread Function estimation for the out-of-focus multispectral compressive imager 219
9.7 Recovered panchromatic images as a function of the undersampling rate 223
9.8 Recovered MS image slices using 25% of the Nyquist-rate measurements 225
9.9 Recovered MS image slices using 50% of the Nyquist-rate measurements 226

List of Statements

1.1	Theorem (WKS Sampling Theorem [36])	8
1.1	Definition (Kolmogorov Complexity [39])	11
1.2	Definition (p-norm)	12
1.3	Definition (k-sparse signal)	12
1.4	Definition ((k, ϑ)-compressible signal)	14
1.5	Definition (Total Variation)	15
1.6	Definition (Mixed (p, q)-norm)	16
1.7	Definition (Jointly k-sparse signals)	17
1.2	Theorem (Singular value decomposition)	17
1.8	Definition (($\breve{\varrho}, \vartheta$)-low rank signals)	18
1.9	Definition (Restricted Isometry Property [12])	24
1.1	Problem (Computation of the k-RIC)	25
1.10	Definition (Coherence [46])	26
1.11	Definition (Sub-Gaussian random variable)	29
1.12	Definition (Sub-Gaussian Random Vector)	29
1.3	Theorem (Singular Values of RsGEs with i.i.d. rows [82, Theorem 5.39])	30
1.4	Theorem (Singular Values of RMEs with i.i.d. entries [85, Theorem 1])	30
1.5	Theorem (RIP of the RsGE [82])	33
1.2	Problem (Linear Least-Squares)	38
1.6	Theorem (Uniqueness of a k-sparse solution [15, 106])	39
1.3	Problem (Sparsest Solution of a Linear System)	39
1.4	Problem (Oracle Least-Squares)	39
1.5	Problem (Basis Pursuit)	40
1.6	Problem (Basis Pursuit with Denoising)	41
1.7	Problem (Least Absolute Shrinkage and Selection Operator)	42
1.8	Problem (Signal Recovery by Convex Optimisation) . .	45
1.7	Theorem (Exact Solution by BP (via coherence) [106])	46

1.8	Theorem (Exact Solution by BP (via the RIP) [64, Theorem 1.1])	46
1.9	Theorem (Stable Recovery by BPDN from Noisy Measurements [64, Theorem 1.2])	47
2.1	Definition (Localisation [2])	57
2.2	Definition (Rakeness (single R.V. case [29]))	58
2.3	Definition (Rakeness (multiple R.V. case))	59
2.1	Problem (Maximum Energy Sensing Matrix Design (i.i.d. rows case))	62
2.1	Proposition (Eigenvalue form of Problem 2.1)	63
2.2	Proposition (Closed-form Solution of Problem 2.1)	64
2.2	Problem (Maximum Energy Sensing Matrix Design (general case))	65
2.3	Proposition (Particular cases of Problem 2.2)	65
2.4	Definition (Anisotropic Random Gaussian Ensemble)	66
2.4	Proposition (Synthesis of an aRGE)	66
2.5	Definition (Anisotropic Random Bernoulli Ensemble)	67
2.5	Proposition (Synthesis of an aRBE by the arcsine law)	67
2.3	Problem (Maximum Energy Sensing Matrix Design (with minimum allocation constraint))	71
4.1	Problem (Maximum Entropy Sensing Matrix Design (deterministic rows case))	96
5.1	Definition (Perturbation Constants [91])	111
5.1	Theorem (Stable Recovery by BPDN in the Presence of Perturbations [91, Theorem 2])	111
6.1	Theorem (Second-class recovery error lower bound (non-asymptotic case))	133
6.2	Theorem (Second-class recovery error lower bound (asymptotic case))	133
7.1	Proposition (Non-Perfect Secrecy of CS [26])	154
7.1	Definition (Asymptotic spherical secrecy)	156
7.2	Proposition (Asymptotic spherical secrecy of i.i.d. RsGE encoding matrices)	157
7.3	Proposition (Rate of convergence with i.i.d. RsGE encoding matrices)	164
8.1	Problem (Subset-Sum Problem)	171
8.1	Proposition (Eve's Known-Plaintext Attack)	172

8.1	Theorem (Expected number of solutions for Eve's KPA)	173
8.2	Theorem (Expected number of solutions for Eve's KPA at a given Hamming distance from the true one)	177
8.2	Problem (γ-cardinality Subset-Sum Problem)	183
8.2	Proposition (Steve's Known-Plaintext Attack)	184
8.3	Theorem (Expected number of solutions for Steve's KPA)	185

List of Acronyms

aBPDN	Analysis BPDN	86
AES	Advanced Encryption Standard	123
APD	Automatic Peak Detection	148
aRBE	Anisotropic Random Bernoulli Ensemble	67
aRGE	Anisotropic Random Gaussian Ensemble	66
ASR	Automatic Speech Recognition	146
AWGN	Additive White Gaussian Noise	74
BPDN	Basis Pursuit with Denoising	41
BP	Basis Pursuit	40
CDF	Cumulative Distribution Function	162
CoSaMP	Compressive Sampling Matching Pursuit	144
CRC	Consecutively Recognised Characters	149
CS	Compressed Sensing	2
DCT	Discrete Cosine Transform	22
DFT	Discrete Fourier Transform	15
DGA	Dense Gaussian Additive	116
DUM	Dense Uniform Multiplicative	116
DWT	Discrete Wavelet Transform	22
ECG	Electrocardiographic Track	57
EMG	Electromyographic Track	57
FPA	Focal Plane Array	208
FP	Fabry-Pérot	202
FT	Fourier Transform	8
GAMP	Generalised Approximate Message-Passing	144

HC	Huffman Coding	81
i.i.d.	independent and identically distributed	27
KKT	Karush-Kuhn-Tucker conditions	64
KLT	Karhunen-Loève Transform	17
KPA	Known-Plaintext Attack	155
KS	Kolmogorov-Smirnov	160
LASSO	Least Absolute Shrinkage and Selection Operator	42
LFSR	Linear Feedback Shift Register	129
LLS	Linear Least-Squares	38
MGF	Moment-Generating function	29
MSB	Most Significant Bit	83
MSE	Mean-Square Error	47
MS	MultiSpectral	18
MU-GAMP	Matrix Uncertainty-GAMP	189
OCR	Optical Character Recognition	149
OLS	Oracle Least-Squares	39
ONB	Orthonormal Basis	13
PCM	Pulse Code-Modulated	80
PDF	Probability Density Function	29
PFE	Partial Fourier Ensemble	28
PHE	Partial Hadamard Ensemble	28
PMF	Probability Mass Function	55
PRNG	Pseudo-Random Number Generator	125
PSD	Positive-Semidefinite	56
PSF	Point Spread Function	203
RBE	Isotropic Random Bernoulli Ensemble	27
RGE	Isotropic Random Gaussian Ensemble	27
k-RIC	Restricted Isometry Constant of order k	24
RIP	Restricted Isometry Property	24
RME	Random Matrix Ensemble	25
R.P.	Random Process	131
RsGE	Random sub-Gaussian Ensemble	28

RTE	Isotropic Random Ternary Ensemble	28
R.V.	Random Vector	29
r.v.	Random Variable	29
SLM	Spatial Light Modulator	215
SNR	Signal-to-Noise Ratio	15
$SPGL_1$	Spectral Projected Gradient for ℓ_1 minimisation	42
SPIHT	Set Partitioning In Hierarchical Trees	82
SSF	Sparse Sign-Flipping	116
SSP	Subset-Sum Problem	171
S-TLS	Sparsity-cognisant Total Least-Squares	189
SVD	Singular Value Decomposition	17
TV	Total Variation	15
UDWT	Undecimated DWT	86
URA	Uniformly Redundant Array	205

List of Symbols

ι	The imaginary unit.
$\mathbb{Z}, \mathbb{Z}_+$	The set of integers and positive integers respectively.
$\mathbb{R}, \mathbb{C}$	The fields of real and complex numbers respectively.
$\mathbb{R}^n, \mathbb{C}^n$	An n-dimensional vector space defined over real or complex numbers respectively.
$\ell_p(\mathbb{Z})$	The Lebesgue space of p-summable sequences.
$L_p(\mathbb{R})$	The Lebesgue space of p-summable functions.
x	Functions, variables or constants (lowercase).
X	Sets (uppercase).
$\mathbf{x}$	Vectors or tensors (boldface, lowercase). $\mathbf{x}$ is a column vector unless specified.
$\mathbf{X}$	Matrices (boldface, uppercase).
x_j, $X_{j,l}$, $x_{j,l,\ldots}$	The j-th element of a vector $\mathbf{x}$, (j,l)-th element of a matrix, $(j,l,\ldots)$-th element of a tensor.
$\{\cdot\}$, $\{\cdot\}_{j=0}^{+\infty}$	Sets and sequences of mathematical entities such as indices, variables, vectors, matrices.
$\lvert \cdot \rvert$	The absolute value of the scalar or vector argument, or the cardinality of a set argument.
$\cdot^c$, $\cdot^\perp$	The complement of the set, or orthogonal complement of the subspace at the argument.

Symbol	Description
$\cdot_T$, $\cdot^S$	The restriction of the vector or set argument to a subset of indices collected in a set T. In the case of matrices, a column submatrix is indicated by $\cdot_T$, a row submatrix is indicated as $\cdot^S$.
$\mathrm{supp}(\cdot)$	The support of the argument over its domain: for a vector, $\{j \in \{0,\ldots,n-1\} : x_j \neq 0\}$; for a function, $\{u \in \mathbb{R} : x(u) \neq 0\}$).
$\mathrm{sign}(\cdot)$	The sign operator on a scalar, vector or matrix argument.
$\mathrm{vec}(\cdot)$	The mapping of a tensor at the argument into a column vector.
$\mathrm{diag}(\cdot)$	The column vector corresponding to the diagonal of a matrix argument, or a diagonal matrix having the vector argument as its diagonal.
$\mathbf{X}^\top$, $\mathbf{X}^*$	Matrix or vector transpose (non-Hermitian and Hermitian, respectively).
$\mathbf{X}^{-1}$, $\mathbf{X}^\dagger$	The inverse and Moore-Penrose pseudoinverse of a matrix, respectively.
$\mathrm{tr}(\mathbf{X})$	The trace of a matrix.
$\det(\mathbf{X})$	The determinant of a matrix.
$\lambda_{\min}, \lambda_{\max}, \lambda_j(\mathbf{X})$	The minimum, maximum and j-th eigenvalue of a matrix (non-increasing order).
$\sigma_{\min}, \sigma_{\max}, \sigma_j(\mathbf{X})$	The minimum, maximum and j-th singular value of a matrix (non-increasing order).
$\|\mathbf{X}\|_2$	The spectral norm of a matrix, $\|\mathbf{X}\|_2 = \sqrt{\sigma_{\max}(\mathbf{X})}$.
$\|\mathbf{X}\|_F$	The Frobenius norm of a matrix, $\|\mathbf{X}\|_F = \sqrt{\mathrm{tr}(\mathbf{X}^*\mathbf{X})}$.
$\|\mathbf{X}\|_\infty$	The entrywise ∞-norm of a matrix, $\|\mathbf{X}\|_\infty = \max|X_{j,l}|$.
$\mathbf{X}\mathbf{Y}$, $\mathbf{X} \circ \mathbf{Y}$, $\mathbf{X} \otimes \mathbf{Y}$	The matrix, Hadamard (element-wise) and Kronecker product of two matrices $\mathbf{X}, \mathbf{Y}$ of suitable dimensions.

List of Symbols

$\stackrel{\text{sed}}{=}, \stackrel{\text{svd}}{=}$	The eigendecomposition of a symmetric matrix $\mathbf{K} \stackrel{\text{sed}}{=} \mathbf{U}\mathbf{\Lambda}\mathbf{U}^*$, and singular value decomposition of a matrix $\mathbf{X} \stackrel{\text{svd}}{=} \mathbf{U}\mathbf{\Sigma}\mathbf{V}^*$.
$\mathbf{I}_n$	The n-dimensional identity matrix.
$\mathbf{0}_n, \mathbf{1}_n$	A column vector or tensor of the specified dimensions, whose elements are all zeros or ones, respectively.
$\hat{\mathbf{x}}, \check{\mathbf{x}}, \tilde{\mathbf{x}}$	An estimate of a vector $\mathbf{x}$, its best approximation with $|\operatorname{supp} \check{\mathbf{x}}| = k$, and a quantisation respectively.
$\mathbb{P}[\cdot], \hat{\mathbb{P}}[\cdot]$	The probability, or empirical probability of the event specified in the argument.
$\xrightarrow[\text{dist.}]{}$	Convergence in distribution.
$f(\cdot), \hat{f}(\cdot)$	The probability density function, or empirical probability density function of the random variable or vector specified in the argument (joint, if not otherwise noted).
$\mathbb{E}[\cdot], \hat{\mathbb{E}}[\cdot]$	The expectation, or empirical expectation of the argument in all its random variables (joint, if not otherwise noted).
$\cdot \vert \cdot$	Conditioning of the first argument with respect to the second one.
$\delta(\cdot)$	The Dirac or Kronecker delta, in the continuous or discrete coordinates specified in the argument respectively.
$\mu_x, \boldsymbol{\mu}_\mathbf{x}$	The mean $\mu_x = \mathbb{E}[x]$ of a random variable x or mean vector $\boldsymbol{\mu}_\mathbf{x} = \mathbb{E}[\mathbf{x}]$ of a random vector $\mathbf{x}$.
$\sigma_x^2, \mathbf{C}_\mathbf{x}, \mathbf{K}_\mathbf{x}$	The variance $\sigma_x^2 = \mathbb{E}[(x-\mu_x)^2]$ of a random variable x, correlation matrix $\mathbf{C}_\mathbf{x} = \mathbb{E}[\mathbf{x}\mathbf{x}^*]$ or covariance matrix $\mathbf{K}_\mathbf{x} = \mathbb{E}[(\mathbf{x}-\boldsymbol{\mu}_\mathbf{x})(\mathbf{x}-\boldsymbol{\mu}_\mathbf{x})^*]$ of a random vector $\mathbf{x}$.
$\mathcal{N}(\boldsymbol{\mu}_\mathbf{x}, \mathbf{K}_\mathbf{x})$	The multivariate Gaussian distribution in $\mathbb{R}^n$ corresponding to $f(\mathbf{x}) = \dfrac{e^{-(\mathbf{x}-\boldsymbol{\mu}_\mathbf{x})^* \mathbf{K}_\mathbf{x}^{-1} (\mathbf{x}-\boldsymbol{\mu}_\mathbf{x})}}{(2\pi)^{\frac{k}{2}} |\det(\mathbf{K}_\mathbf{x})|^{\frac{1}{2}}}$.

$\mathcal{U}(a,b)$ The univariate uniform distribution defined on the interval $[a,b] \in \mathbb{R}$.

INTRODUCTION

THE ubiquitous use of sensor data in modern engineering applications highlights a large number of cases in which a considerable amount of resources such as acquisition time, power and analog or optical hardware are spent to implement relatively high-resolution sensing interfaces that provide a lossless discrete representation of a continuously varying quantity of interest.

However, in many cases this representation shows an intrinsic *redundancy* that is effectively exploited by data compression algorithms to reduce the length of the binary string that encodes the salient information content. This redundancy can be seen as a manifestation of an underlying stationarity, repeatability, or in general terms *structure* that is found in natural signals, with the noteworthy exception of noise.

Among the methods used to highlight such a structure, the application of linear transforms is widely recognised as a mean to describe a signal in a domain where it is accurately represented by a relatively small number of non-zero coefficients. This basic concept of a *sparse signal model* leads to understanding that, even prior to its acquisition, a signal could be accurately described by capturing these coefficients, whose number or *sparsity* serves as an index for a signal's *complexity*.

In the last decade the idea that the sensing method used to acquire a signal should be strictly balanced to this notion of complexity has gained overwhelming attention under the name of *Compressed Sensing (CS)* [11]. This framework for signal acquisition and processing rephrases the problem of sampling as a *dimensionality reduction*, *i.e.*, a linear operator that maps a full-resolution representation of the signal of interest into an *undersampled set of measurements*. Some fundamental theoretical results [12–16] showed that, while such a dimensionality reduction generally implies inevitable information loss, when a signal is sufficiently *sparse* it may also be recovered perfectly from its undersampled measurements by means of a suitable *signal recovery algorithm*.

Even more surprisingly, a *non-adaptive, universal* approach to this dimensionality reduction amounts to applying a suitable random matrix [17, 18] in the signal domain. In this view CS could be summarised as the fascinating idea of *sensing reality by means of a random code*.

While the rich mathematical foundations of this theory are now consolidated, "second-generation" issues arise in how this paradigm should be implemented to address the difficulties and open challenges of modern scientific data acquisition interfaces, as well as how it could improve upon the state-of-the-art of engineering applications. In fact, while theoretical results on CS proliferate in the literature, the actual implementation of this technique has been slow and cautious for a number of reasons.

The first reason is that the application of a random code in the analog or optical domain often leads to non-trivial sensing interface designs; these have to be justified by remarkable undersampling regimes paired with accurate recovery of the acquired signal (*e.g.*, [19, 20]), or equivalently by the possibility of attaining otherwise unachievable high resolution with existing technologies, *i.e.*, the so-called *super-resolution*.

The second reason is that a sparse signal model *per se* is often insufficient to guarantee signal recovery with acceptable performances; in fact, there is much more structure to leverage in natural signals than sparsity, which is by itself a strictly deterministic signal model.

Fortunately, the literature does not fall short of *model-based* extensions (see, *e.g.*, [21–23]) of sparsity to improve signal recovery. However, it is here advocated that with a change in the signal model a modification of the random code should also follow, *i.e.*, that *adaptive* sensing operator designs can lead to higher undersampling rates or equivalently to more accurate signal recoveries.

The third reason is that CS is effectively applied only when it truly represents a low-complexity alternative to reducing the resources spent in tasks that are related to the manipulation and transmission of sensor data. In fact, while some negative results [24, 25] have suggested the contrary, an embodiment of CS to *encode* a signal after its acquisition holds some value as a non-adaptive technique for lossy signal compression using an extremely simple multiplierless digital architecture. In addition, early contributions [12, 26] envision the possibility of using it as a means to provide *encryption* capabilities by using random codes, thus integrating security properties directly in the sensing process.

Contributions and Outline

This thesis aims at addressing these challenges by means of methods and matrix designs that allow the introduction of specific features in the random code. In particular, we move in the two directions of *efficient* and *secure* Compressed Sensing, where the first direction indicates the possibility of finding adaptive matrix designs that efficiently allow the achievement of improved undersampling rates when compared with those attained by standard CS. On the other hand, the second direction aims at using CS as a method to provide a low-complexity encryption that is integrated in its linear encoding stage.

The general procedure adopted in this work involves the development of theory, methods (supported by accurate and thorough numerical simulation and proof sketches) and heuristics on (i) how to obtain signal-adaptive sensing matrix designs and (ii) what security properties can be provided in a secure communication scheme based on CS.

In particular, this work is organised as follows:

- **Chapter 1** presents an overview of Compressed Sensing and its fundamentals, with the aim of providing a standalone background to this thesis. While no novel material is presented in this Chapter, its role is essential in agreeing on the notation and concepts addressed by the author in the rest of this manuscript.

- **Part I** generally addresses the adaptation of sensing matrices to the task of acquiring *localised signals, i.e.,* signals with a strongly non-white correlation matrix. The objective of this Part is promoting proxies for the concept of "information extraction" in the presence of a localised signal. **Chapter 2** presents random sensing matrix designs based on maximising the compressive measurements' average energy, a proxy that is enforced by the concept of *rakeness* as initially developed in [27–29]; the author has rephrased this maximisation in terms of localisation, as formally introduced in [2], and has provided substantial evidence that *rakeness* operates in absence of noise, on exactly sparse signals and in a fashion that is linked to localisation. **Chapter 3** proposes an application of maximum energy sensing matrix designs to the task of encoding biosignals [3, 9], showing how a code rate reduction is possible by means of such adaptive designs with respect to standard sensing (or *encoding*) matrices, approaching the rates of higher-complexity, higher-performances algorithms. **Chapter 4** addresses a different perspective in sensing matrix design, *i.e.,* the problem of finding an optimal subset of deterministic vectors to form a random matrix by selection in this design space; the rationale adopted here is a maximisation of the compressive measurements' *differential entropy* as a proxy for the amount of information embedded into them. This is shown to correspond to an optimisation problem whose solution yields a sensing matrix design which outperforms randomly constructed matrices in the same design space [4].

- **Part II** addresses an encryption protocol based on CS and its sensitivity to matrix perturbations. **Chapter 5** discusses this aspect of sensitivity, and provides a method to evaluate the average performances of CS under matrix perturbations [8], *e.g.*, caused by missing information on the sensing matrix. **Chapter 6** exploits this decoder-side sensitivity to introduce a multiclass encryption scheme [5] by which different users are provably capable of recovering a signal only up to a prescribed quality level, thus creating different classes of access to the information content encoded by the compressive measurements; this is seen by developing some recovery error bounds for *lower-class* users. The matter of assessing the security of such a simple encryption scheme is addressed separately in **Chapter 7** and **Chapter 8**. The former is concerned with providing evidence on a statistical cryptanalysis of CS as provided by a broad class of random matrices, showing how the requirement for perfect secrecy can be relaxed (both asymptotically and non-asymptotically) into a notion of secrecy conditioned only on the power or energy of the plaintext. Chapter 8 focuses on computational cryptanalysis techniques in the form of Known-Plaintext Attacks, *i.e.*, once the attacker is provided with one plaintext-ciphertext pair. This analysis is borrowed from [6]. We will show by a theoretical analysis (accurately matched by numerical evidence) that the number of sensing matrix configurations corresponding to such a pair is very large, and in absence of side-information a decision in favour of a particular solution cannot be made. This fact holds, with different solution sets, for both eavesdroppers and non-perfectly informed, lower-class users performing an attack to upgrade their recovery quality.

▶ **Part III**, solely comprised of **Chapter 9**, is a treatment on the design of a multispectral compressive imager, as part of the collaboration of the author with the Integrated Imagers Division of IMEC, Belgium. The designed imager extends the concept of random convolution [30–32] to the case of multispectral imaging, with the aim of providing a reduction in the active pixel count of a specific class of multispectral CMOS sensors [7]. An overview of the challenges in the realisation of a prototype for this multispectral imaging architecture is presented, along with the problem of completing its mathematical model by estimating its Point Spread Function, that is crucial in improving the attained recovery quality. Early recovery results are presented and leave room for future improvement, in pursuit of a novel multispectral imager design as a candidate application for Compressed Sensing.

ELEMENTS OF SPARSE SIGNAL MODELS AND COMPRESSED SENSING

THE problem of accurately acquiring and digitising a *continuous signal*, *i.e.*, a source of information in the form of a function $x(u) \in \mathbb{R}$ defined, *e.g.*, on a one-dimensional *continuum* $u \in \mathbb{R}$ (such as time) subtends fundamental challenges in empirical sciences, where experiments that gather knowledge about physical phenomena require an abundance of measurements. Clearly, these must be acquired by accurate interfaces between the analog and digital domain, where most of the information processing, storage and transmission occurs. Such operations are carried out on *samples* $x(u_j) \in \mathbb{R}$ that represent $x(u)$ in a sequence $\{x(u_j)\}_{j\in\mathbb{Z}}$ defined on a discrete sampling grid $\{u_j\}_{j\in\mathbb{Z}}, u_j \in \mathbb{R}$. To complete a digital representation of the sequence, each sample will be mapped to a symbol in some finite set or *alphabet* $\mathcal{X}$ that allows a *quantisation*[1] of the continuous-valued samples as $\tilde{x}(u_j) = \mathcal{Q}\left[x(u_j)\right], \mathcal{Q} : \mathbb{R} \to \mathcal{X}$. Very generally, this pair of operations may be denoted as the composition of a *sensing operator*[2] $\mathcal{A} : L_2(\mathbb{R}) \to \ell_2(\mathbb{Z})$ depending on the chosen sampling method and grid, followed by the scalar quantiser $\mathcal{Q}$. These few sentences open

---

[1] Only scalar quantisers [33] are assumed by this definition. Vector quantisers [34] operating on the sequence are normally applied after a first scalar quantisation of a continuous signal, and lie outside the topics discussed in this thesis.

[2] The concepts of Lebesgue spaces, p-summable functions and sequences are not reviewed here but found in [35, Chap. 3].

a number of problems such as: what properties should the sensing operator $\mathcal{A}$ be endowed with in order to determine unequivocally $x(u)$ by its samples? And what quantiser $\mathcal{Q}$ allows the minimum loss of information *with respect to (w.r.t.)* continuous-valued samples? The following result of Whittaker-Kotel'nikov-Shannon (WKS) is a classic, broadly applicable answer to the first question.

Theorem 1.1 (*WKS Sampling Theorem [36]*). *Let $x(u) \in L_2(\mathbb{R})$ have Fourier Transform (FT)*

$$\mathcal{F} : L_2(\mathbb{R}) \to L_2(\mathbb{R}), \mathcal{F}[x](v) = \int_{\mathbb{R}} x(u) e^{-\imath 2\pi u v} du$$

which is compactly supported, *i.e.*, $\operatorname{supp}_{v \in \mathbb{R}}(|\mathcal{F}[x](v)|) \subseteq [-\beta, \beta]$ *for some finite $\beta > 0$. Consider a sequence of samples uniformly and equally spaced by $\tau > 0$, i.e.,*

$$\{x(u_j)\}_{j \in \mathbb{Z}} = \{x(j\tau)\}_{j \in \mathbb{Z}}$$

If $\tau \leq \frac{1}{2\beta}$ then

$$x(u) = \sum_{j \in \mathbb{Z}} x(j\tau) \operatorname{sinc}\left(\frac{u}{\tau} - j\right) \quad (1.1)$$

Thus, a sufficient condition to ensure that a signal is perfectly represented by its samples is that the sensing operator $\mathcal{A}$ must not introduce *aliasing* (*i.e.*, superposition in the FT-domain and consequent information loss), therefore implying lossless recovery of $x(u)$ from its samples by the interpolation formula (1.1). Most modern signal acquisition interfaces therefore operate at or above the *Nyquist rate* $f_N = 2\beta$, so that a finite-length[3] observation in $u \in [0, \frac{n}{2\beta}]$ maps to at least n samples collected in a vector $\mathbf{x} = \begin{bmatrix} x(u_0) & x(u_1) & \cdots & x(u_{n-1}) \end{bmatrix}^\top = \mathcal{A}[x(u)], u \in [0, \frac{n}{2\beta}]$. We let this *discrete signal* $\mathbf{x} \in \mathbb{R}^n$ be the Nyquist-rate representation[4] of $x(u)$, denoting its quantised version as $\tilde{\mathbf{x}} = \mathcal{Q}[\mathbf{x}] \in \mathcal{X}^n$. When more prior information is

---
[3]The careful discussion of a more general sampling theorem is found in [37].
[4]$\mathbf{x}$ is considered a perfect representation of the continuous signal by Theorem 1.1. Continuous signals and transforms are not used for a large part of this thesis. $\mathbf{x}$ is

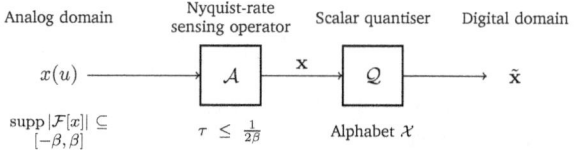

Figure 1.1: A standard Nyquist-rate sensing scheme.

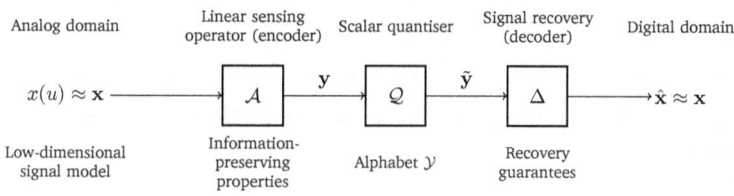

Figure 1.2: A generic compressed sensing scheme

available in the form of a *signal model* alternatives to the conditions of Theorem 1.1 can be envisioned, by which a smaller number of samples could suffice to recover a nearly-lossless approximation of the original signal. Model-based sensing methods that exploit additional priors on x to select an efficient operator $\mathcal{A}$ have been the subject of intense research in the last decade under the common name of CS [11,38]. The elements that interact in this mathematical framework are summarised as follows (see Fig. 1.2, 1.3):

- **Low-Dimensional Signal Models:** the definition of a sensing method requires a prior model describing how the information content in x is distributed (*e.g.*, in a suitable transform domain). In particular, the classic theory of CS leverages deterministic[5] low-dimensional signal models, *i.e.*, signal representations by which $\mathbf{x} \in \mathbb{R}^n$ truly lies in a subspace with significantly smaller dimensionality;

- **Sensing Operators and Properties:** a linear sensing operator $\mathcal{A}: \mathbb{R}^n \to \mathbb{R}^m, m < n$ is used to acquire x, performing a

---

referred to as the *acquired signal* or simply the *signal*.

[5]In the sense that no explicit hypothesis is made *a priori* on the probability distribution of x.

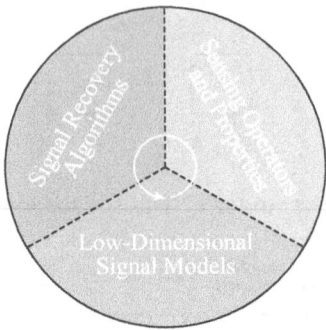

Figure 1.3: The mathematical elements that define the framework of compressed sensing.

dimensionality reduction that maps it to a *measurement vector* $\mathbf{y} = \mathcal{A}(\mathbf{x}) \in \mathbb{R}^m$. The aim of this operation is minimising the resources spent by the sensing operator. This setting is similar to many other inverse problems if not for the fact that, since $m < n$, the problem of recovering $\mathbf{x}$ from $\mathbf{y}$ is ill-posed. In spite of this, the design of $\mathcal{A}$ will verify some *information-preserving properties* related to the prior model on $\mathbf{x}$ and ensuring its bounded-error recoverability from $\mathbf{y}$;

▶ **Signal Recovery Algorithms**: if the above properties are verified, theoretical guarantees will ensure that an estimate $\hat{\mathbf{x}}$ of $\mathbf{x}$ can be inferred from $\mathbf{y}$ by means of suitable signal recovery algorithms (Δ in Fig. 1.2) even when the measurements are subject to additive noise, such as that introduced by a quantiser ($\mathcal{Q}$ in Fig. 1.2) yielding $\tilde{\mathbf{y}}$ or, e.g., when the sensing operator is perturbed. Signal recovery is here reviewed in terms of *convex optimisation problems* that explicitly enforce a low-dimensional model in the approximation of $\mathbf{x}$ and can be solved by existing algorithms for this problem class. In addition, fast *greedy algorithms* exist to minimise the computational complexity of signal recovery by accepting a heuristic solution.

Each of the highlighted parts is comprised of a dense, rapidly evolving literature of which only the fundamentals are reviewed in the following

Sections as a background to this thesis.

We remark that the purpose of CS is to provide a solid mathematical background for the design of signal acquisition methods, *i.e.*, describing new sensing operators and their corresponding signal recovery procedures prior to any implementation-level consideration. The emphasis of this review Chapter is on taking a user's perspective and understanding the relationships between a plethora of existing mathematical tools; for this reason, we do not delve at this level into either technical proofs (well documented in the references) or specific analog-to-digital or digital-to-digital applications of CS that will naturally, albeit non-trivially follow.

1.1 Low-Dimensional Signal Models

Among the possible *a priori* hypotheses that can be made on a signal, the ones analysed here share a general and common notion of low *complexity*, whose motivation may be found in the following definition.

Definition 1.1 *(Kolmogorov Complexity [39])*. Let Υ be a *universal computer* [40] that processes any finite binary string $\tilde{s} \in \{0,1\}^{l_{\tilde{s}}}$ of arbitrary length $l_{\tilde{s}}$ providing a description of $\tilde{x} \in \mathcal{X}^n$. The *Kolmogorov complexity* of $\tilde{x}$ is

$$\min_{\tilde{s}} l_{\tilde{s}} \text{ s.t. } \Upsilon(\tilde{s}) = \tilde{x}$$

According to this statement, $\tilde{x}$ is only as complex as its shortest binary description that programs Υ to produce it at the output. This early, very general and incomputable definition (clearly different from Shannon's practical notion of *entropy* as a probabilistic measure of information [41]) expresses the idea that the information content in an object decreases as its intrinsic, deterministic structure (repeatability, redundancy) increases.

In the case of $\mathbf{x} \in \mathbb{R}^n$, we here review deterministic models in which a signal is described with the least possible number of dimensions[6], in analogy with Definition 1.1. Deterministic *sparse* and *low-rank* models are here introduced to find low-dimensional descriptions by which $\mathbf{x}$ is accurately and succinctly represented. Many other methods exist to exploit such structures in the approximation of a signal; their review is left to [23, 43–45] and references therein.

1.1.1 Sparse Signal Models

Firstly, we recall the notion and notation of *p-norm*.

Definition 1.2 (*p-norm*). The *p-norm* (also ℓ_p-norm) of a vector $\mathbf{x} \in \mathbb{R}^n$ is defined as

$$\|\mathbf{x}\|_p = \left(\sum_{j=0}^{n-1} |x_j|^p \right)^{\frac{1}{p}} \quad (1.2)$$

Note that for

- $p = \infty$, $\|\mathbf{x}\|_\infty = \max_{j \in \{0,\ldots,n-1\}} |x_j|$;
- $p = 0$, $\|\mathbf{x}\|_0 = \left| \mathrm{supp}_{j \in \{0,\ldots,n-1\}}(\mathbf{x}) \right|$ is a *pseudo*-norm;
- $p \geq 1$, (1.2) is a *convex* function (see [35, Def. 3.1]);
- $p \in [0, 1)$, (1.2) is a *non-convex* function.

We now proceed to reviewing sparse signal models.

Sparse Signals

A straightforward model to define a low-complexity description of $\mathbf{x}$ is based on sparsity, as summarised by the following definition.

Definition 1.3 (*k-sparse signal*). Consider a *sparsity basis* or *dictionary* of $\mathbb{R}^n$ whose vectors (or *atoms*) $\{\mathbf{d}_j\}_{j=0}^{p-1}$ are collected in a matrix

---

[6]Other noteworthy connections can be made with the statistical principle of minimum description length [42].

$\mathbf{D} = \begin{bmatrix} \mathbf{d}_0 & \cdots & \mathbf{d}_{p-1} \end{bmatrix} \in \mathbb{R}^{n \times p}, p \geq n$.

A signal $\mathbf{x} \in \mathbb{R}^n$ is *k-sparse* w.r.t. $\mathbf{D}$ if there exists a vector $\mathbf{s} \in \mathbb{R}^p$ for which $\mathbf{x} = \mathbf{D}\mathbf{s}$ and $\|\mathbf{s}\|_0 = k$ with $k \ll n$.

Thus, $\mathbf{x}$ is formed by a linear combination of some $\{\mathbf{d}_j\}_{j \in T}$, $T = \operatorname{supp}(\mathbf{s}) = \{j \in \{0, \ldots, p-1\} : s_j \neq 0\}$ as $\mathbf{x} = \mathbf{D}_T \mathbf{s}_T$. Equivalently, $\mathbf{x}$ is described by $\mathbf{s} \in \Sigma_k$, $\Sigma_k \subset \mathbb{R}^n$ with

$$\Sigma_k = \bigcup_{T \subseteq \{0,\ldots,p-1\}, |T|=k} U_T, \quad U_T \subseteq \mathbb{R}^k$$

denoting the union of all $\binom{p}{k}$ k-dimensional canonical subspaces U_T. Sparsity is therefore a measure of a signal's complexity by the cardinality of its description w.r.t. a suitable transform, in a fashion similar to Definition 1.1.

A distinction must then be made between sparsity in a *basis* ($p = n$) or an *overcomplete dictionary* ($p > n$). In the former case and assuming an *Orthonormal Basis (ONB)* the corresponding $\mathbf{D}$ is *unitary* ($\mathbf{D}^{-1} = \mathbf{D}^*$) and the relation $\mathbf{s} = \mathbf{D}^* \mathbf{x}$ is invertible. In the latter case the inverse of $\mathbf{D}$ is undefined, and $\mathbf{x} = \mathbf{D}\mathbf{s}$ is an underdetermined system of equations (*i.e.*, a non-injective linear transformation) with infinite solutions $\mathbf{s}$ among which the sparsest is sought [46].

This problem can be tackled by using the prior $\|\mathbf{s}\|_0 = k$ to perform a combinatorial search w.r.t. T, testing whether

$$\mathbf{s} = \begin{cases} \mathbf{s}_T = \mathbf{D}_T^\dagger \mathbf{x}, & |T| = k \\ \mathbf{s}_{T^c} = \mathbf{0}_{p-k}, & T^c = \{0, \ldots, p-1\} \setminus T \end{cases} \quad (1.3)$$

is a solution to $\mathbf{x} = \mathbf{D}\mathbf{s}$ for each of $\binom{p}{k}$ possible supports, noting that $\mathbf{x} = \mathbf{D}_T \mathbf{s}_T$ is an overdetermined system of equations (*i.e.*, a non-surjective linear transformation). However (*i*) unsurprisingly this test leads to an NP-hard problem [47] and (*ii*) even if such a computational complexity was affordable, uniqueness conditions for sparse solutions would still have to be proved. Since this problem is analogous to those imposed by CS, we postpone to Section 1.3 the discussion of when and how the sparsest $\mathbf{s}$ may be found by polynomial-complexity algorithms.

Admitting that they exist, sparsity in such overcomplete dictionaries often leads to values of k smaller than with an ONB.

This brief summary of sparse signal modelling is carried out from the so-called *synthesis* perspective, referring to the *synthesis transform* $\mathbf{x} = \mathbf{D}\mathbf{s}$ of a pair $(\mathbf{D}^\star, \mathbf{D})$; $\mathbf{D}^\star$ is the corresponding *analysis transform* such that $\mathbf{s}^\star = \mathbf{D}^\star \mathbf{x}$ (in general, $\mathbf{s} \neq \mathbf{s}^\star$). When overcomplete dictionaries are considered, searching for either sparse $\mathbf{s}$ or $\mathbf{s}^\star$ leads to very different approximation performances, as noted in the literature [45, 48–50]. The *analysis* perspective will be explicitly used and expanded where needed, along with specifications on how the above $(\mathbf{D}^\star, \mathbf{D})$ are chosen.

Compressible Signals

While exactly k-sparse representations of $\mathbf{x}$ are infrequently found when sampling physical quantities, it is very often verified that an accurate k-sparse approximation can be obtained from a non-sparse $\mathbf{s}$ by *thresholding* all but the $k \ll p$ elements with the largest absolute value, i.e., sorting the indices $\{j_0, \ldots, j_{p-1}\}$ so that $|s_{j_0}| \geq |s_{j_1}| \geq \ldots \geq |s_{j_{p-1}}|$ and letting

$$\check{\mathbf{s}} = \begin{cases} \check{\mathbf{s}}_T = \mathbf{s}_T, & T = \{j_0, \ldots, j_{k-1}\} \\ \check{\mathbf{s}}_{T^c} = \mathbf{0}_{p-k}, & T = \{0, \ldots, p-1\} \setminus \{j_0, \ldots, j_{k-1}\} \end{cases} \quad (1.4)$$

More generally, and among many alternatives [43, 51], the following definition indicates when a signal is *compressible*, i.e., effectively approximated with high accuracy by a k-sparse representation.

Definition 1.4 ((k, ϑ)-*compressible signal*). A signal $\mathbf{x} \in \mathbb{R}^n$ is (k, ϑ)-compressible w.r.t. $\mathbf{D}$ if the *best k-sparse approximation*

$$\check{\mathbf{s}} = \operatorname*{argmin}_{\boldsymbol{\xi} \in \Sigma_k} \|\mathbf{x} - \mathbf{D}\boldsymbol{\xi}\|_2 \quad (1.5)$$

is so that

$$\|\mathbf{x} - \mathbf{D}\check{\mathbf{s}}\|_2 \leq \vartheta \|\mathbf{x}\|_2, \ \vartheta \geq 0$$

This definition indicates that the resulting approximation *Signal-to-Noise Ratio (SNR)*,

$$\text{SNR}_{\check{\mathbf{s}},\mathbf{s}} = 20 \log_{10}\left(\frac{\|\mathbf{s}\|_2}{\|\check{\mathbf{s}} - \mathbf{s}\|_2}\right)$$

is as large as set by ϑ. In addition, it must be noted that the same SNR is observed in the signal domain only when $\mathbf{D}$ is an ONB. On the contrary, when $\mathbf{D}$ is an overcomplete dictionary the $\text{SNR}_{\check{\mathbf{x}},\mathbf{x}}$ of $\check{\mathbf{x}} = \mathbf{D}\check{\mathbf{s}}$ will have to be calculated; moreover, in that case (1.4) will not necessarily solve (1.5).

This compressibility property has been observed, *e.g.*, in the representation of natural images by means of a suitable Discrete Wavelet Transform (DWT, see [52] for a comprehensive review) whose transform-domain coefficients exhibit a rapid decay in absolute value, *i.e.*, a relatively small number of them accurately describe the salient information content in an image. For such non-sparse signals, a notion of *power-law decay* may also be used to quantify compressibility (see the review in [43]).

Total Variation-Sparse Signals

A common alternative to orthonormal transforms such as the *Discrete Fourier Transform (DFT)* or DWT is the *Total Variation (TV)* of x, defined as follows in the general case of a *three-dimensional (3D)* tensor.

Definition 1.5 *(Total Variation)*. Let $\mathbf{x} \in \mathbb{R}^{n \times m \times q}$ be a 3D tensor. We define *anisotropic total variation* as

$$\|\mathbf{x}\|_{\text{TV}} = \sum_{i=0}^{n-1}\sum_{j=0}^{m-1}\sum_{l=0}^{q-1} |x_{i+1,j,l} - x_{i,j,l}| + \\ + |x_{i,j+1,l} - x_{i,j,l}| + |x_{i,j,l+1} - x_{i,j,l}| \quad (1.6)$$

and *isotropic total variation* as

$$\|\mathbf{x}\|_{\text{TVi}} = \sum_{i=0}^{n-1}\sum_{j=0}^{m-1}\sum_{l=0}^{q-1}(|x_{i+1,j,l}-x_{i,j,l}|^2+$$

$$+|x_{i,j+1,l}-x_{i,j,l}|^2+|x_{i,j,l+1}-x_{i,j,l}|^2)^{\frac{1}{2}} \quad (1.7)$$

This semi-norm[7] measures the piecewise-smoothness of x in its domain; in this model, a signal can be regarded as sparse w.r.t. TV if it is piecewise-constant (in the case of images, this is the so-called "cartoon" model). Clearly, when *two-dimensional (2D)* images or one-dimensional signals are considered (1.6), (1.7) will drop the terms related to the unneeded dimensions. Both (1.6) and (1.7) are widely used as cost functions for signal recovery since their introduction for image denoising applications [53].

Joint-Sparse Signals

The simplest case of a sparse signal model capable of leveraging the dependence between multiple *instances* of x, collected in the columns of a matrix $\mathbf{X} = \begin{bmatrix} \mathbf{x}_0 & \mathbf{x}_1 & \cdots & \mathbf{x}_{w-1} \end{bmatrix}$, $\mathbf{X} \in \mathbb{R}^{n \times w}$, is that of *joint-sparsity*. The following definition is commonly employed in joint-sparse models.

Definition 1.6 *(Mixed (p,q)-norm)*. The *mixed (p,q)-norm* of a matrix $\mathbf{X} \in \mathbb{R}^{n \times w}$ is defined as

$$\|\mathbf{X}\|_{p,q} = \left(\sum_{j=0}^{n-1}\left(\sum_{l=0}^{w-1}|X|_{j,l}^p\right)^{\frac{q}{p}}\right)^{\frac{1}{q}}$$

In particular, for $p=2, q=0$ the pseudo-norm

$$\|\mathbf{X}\|_{2,0} = \left|\operatorname*{supp}_{j \in \{0,\ldots,n-1\}}\left(\sqrt{\sum_{l=0}^{w-1}|X|_{j,l}^2}\right)\right|$$

---

[7] A basic property of norms is not verified since $\|\mathbf{x}\|_{\text{TV}} = 0 \not\Rightarrow \mathbf{x} = \mathbf{0}_{n \times m \times q}$.

is non-convex.

The notion of joint-sparsity recalled here is also known in the literature as JSM-2 [54, 55].

Definition 1.7 *(Jointly k-sparse signals)*. A matrix of signal instances $\mathbf{X} \in \mathbb{R}^{n \times w}$ is *jointly k-sparse* w.r.t. $\mathbf{D}$ if there exists a set of sparse vectors $\mathbf{S} \in \mathbb{R}^{p \times w}$ for which $\mathbf{X} = \mathbf{D}\mathbf{S}$ and $\|\mathbf{S}\|_{2,0} = k$.

Thus, it is assumed that a common k-sparse support is shared between the representations of multiple instances of the same signal, either due to their similarity or to physically distinct sensors simultaneously acquiring the same quantity in a distributed scheme. Extensions of this model are, *e.g.*, reviewed in [44].

1.1.2 Low-Rank Signal Models

The previously introduced model of joint-sparsity assumed a particular structure in $\mathbf{X} = \mathbf{D}\mathbf{S}$ by promoting the similarity between the column supports of $\mathbf{S}$. More generally, factorising a matrix of observations as a product of structured matrices is ubiquitous in data analysis. *Singular Value Decomposition (SVD)*[8] is perhaps the most basic of such methods [56–58].

Theorem 1.2 *(Singular value decomposition)*. Let $\mathbf{X} \in \mathbb{R}^{n \times w}$ be a rectangular matrix. $\mathbf{X}$ admits a *singular value decomposition* of the form

$$\mathbf{X} \stackrel{svd}{=} \mathbf{U}\mathbf{\Sigma}\mathbf{V}^* = \sum_{j=0}^{\varrho-1} \mathbf{u}_j \sigma_j(\mathbf{X}) \mathbf{v}_j^*$$

where

- $\varrho = \text{rank}(\mathbf{X}) \leq \min\{n, w\}$;
- $\mathbf{\Sigma} \in \mathbb{R}^{n \times w}$ is a rectangular diagonal matrix collecting the

---

[8]It is also widely known as principal components analysis or *Karhunen-Loève Transform (KLT)*, where SVD is the method used to compute this transform on statistical data samples.

singular values $\sigma_j(\mathbf{X}) = \sqrt{\lambda_j(\mathbf{X}^*\mathbf{X})} = \sqrt{\lambda_j(\mathbf{X}\mathbf{X}^*)} \geq 0$ in a non-increasing sequence $\{\sigma_j(\mathbf{X})\}_{j=0}^{\min\{n,w\}-1}$;

- $\mathbf{U} \in \mathbb{R}^{n \times n}$, $\mathbf{V} \in \mathbb{R}^{w \times w}$ are unitary matrices collecting the left- and right-*singular vectors* of $\mathbf{X}$, *i.e.*, $\{\mathbf{u}_j\}_{j=0}^{n-1}$ and $\{\mathbf{v}_j\}_{j=0}^{w-1}$ such that $\mathbf{X}\mathbf{v}_j = \sigma_j(\mathbf{X})\mathbf{u}_j$ and $\mathbf{u}_j^*\mathbf{X} = \sigma_j(\mathbf{X})\mathbf{v}_j$ respectively.

This statement follows by application of the spectral theorem [59, Chap. 1] to the symmetric matrices $\mathbf{X}\mathbf{X}^*$ and $\mathbf{X}^*\mathbf{X}$. In addition, a consequent result of Eckart and Young [60] proves that when $\mathbf{X}$ is a ϱ-rank matrix its best $\check{\varrho}$-rank approximation $\check{\mathbf{X}}$ (in the least-squares sense[9]) with $\check{\varrho} < \varrho$ is $\check{\mathbf{X}} \stackrel{\mathrm{svd}}{=} \mathbf{U}\check{\boldsymbol{\Sigma}}\mathbf{V}^*$, where $\check{\boldsymbol{\Sigma}}$ collects the $\check{\varrho}$ largest singular values in $\boldsymbol{\Sigma}$, setting the remaining $\varrho - \check{\varrho}$ to 0. A definition of such a low-rank approximation for multiple signal instances follows.

Definition 1.8 (($\check{\varrho}, \vartheta$)-*low rank signals*). A matrix of signal instances $\mathbf{X}$ has a ($\check{\varrho}, \vartheta$)-*low rank* if its *best $\check{\varrho}$-rank approximation*,

$$\check{\mathbf{X}} = \operatorname*{argmin}_{\substack{\boldsymbol{\Xi} \in \mathbb{R}^{n \times q} \\ \check{\varrho} = \mathrm{rank}(\boldsymbol{\Xi})}} \|\boldsymbol{\Xi} - \mathbf{X}\|_F \stackrel{\mathrm{svd}}{=} \mathbf{U}\check{\boldsymbol{\Sigma}}\mathbf{V}^* = \sum_{j=0}^{\check{\varrho}-1} \mathbf{u}_j \sigma_j(\mathbf{X}) \mathbf{v}_j^*$$

is such that

$$\|\check{\mathbf{X}} - \mathbf{X}\|_F \leq \vartheta \|\mathbf{X}\|_F, \vartheta \geq 0$$

Thus, the resulting $\mathrm{SNR}_{\check{\mathbf{X}}, \mathbf{X}} = 20\log_{10} \frac{\|\mathbf{X}\|_F}{\|\check{\mathbf{X}} - \mathbf{X}\|_F}$ is as large as set by $\check{\varrho}$.

1.1.3 An Example: Low-Dimensional Models of Multispectral Data

As a practical example of sparse signal modelling we here consider the case of a *MultiSpectral (MS)* image or cube, *i.e.*, a 3D tensor in which each *voxel* represents a light intensity corresponding to two spatial coordinates and a specific wavelength in the electromagnetic spectrum. Such data volumes have very large dimensionality, and carry highly

---

[9] Minimising the sum of square residuals between an approximation or estimate and the actual quantity; in this case, $\|\check{\mathbf{X}} - \mathbf{X}\|_F$.

(a) RGB colour image of the scene.

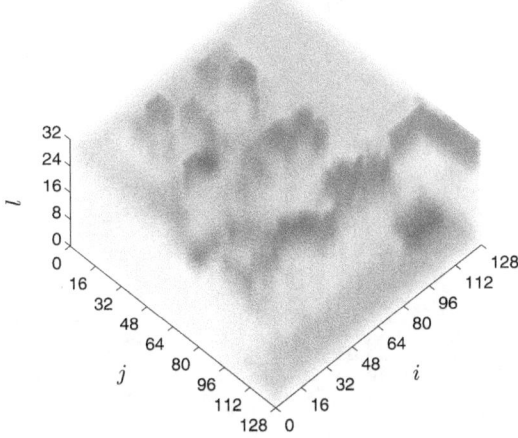

(b) 3D visualisation of $\mathbf{x}$. The wavelength indices are mapped to the matching colour; the voxel transparency is proportional to $x_{i,j,l}$. Structure in the wavelength domain can be visually appreciated in the data volume.

Figure 1.4: The "Fresh Fruit" MS sample image.

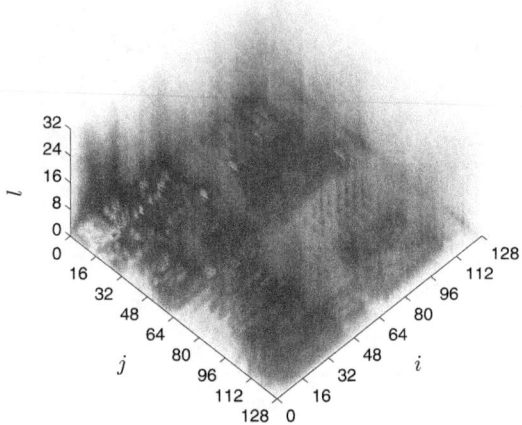

(a) 3D visualisation of **s**. The voxel transparency and colour are proportional to $|s_{i,j,l}|$.

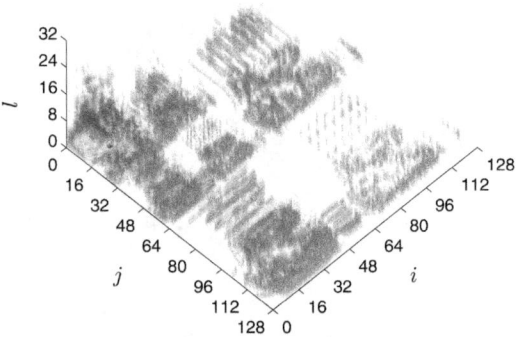

(b) 3D visualisation of $\tilde{\mathbf{s}}$ obtained as the best $k = 19073$-sparse approximation so that $\mathrm{SNR}_{\tilde{\mathbf{s}},\mathbf{s}} = 40\,\mathrm{dB}$.

Figure 1.5: Representation of a MS image on a sparsity basis **D** obtained by Kronecker product of a 2D Haar DWT and a DCT basis.

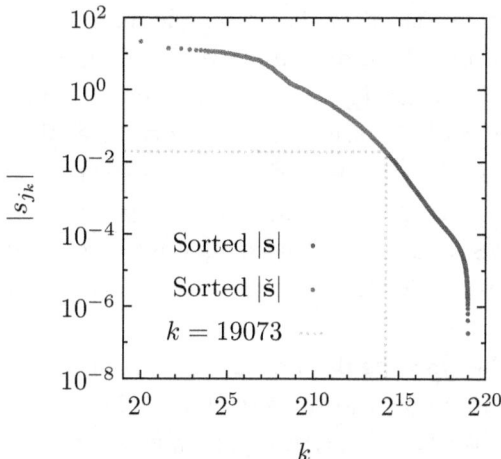

(a) Decay of **s** and its best k-sparse approximation $\check{\mathbf{s}}$ ($\log_2 - \log_{10}$ scale).

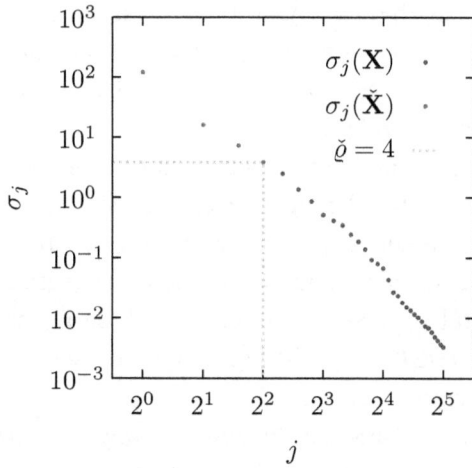

(b) Singular values of **X** and its best $\check{\varrho}$-rank approximation $\check{\mathbf{X}}$ ($\log_2 - \log_{10}$ scale).

Figure 1.6: Evaluation of the compressible and low-rank behaviour of a MS image.

structured information suitable for material and chemical compound identification purposes.

We here consider the "Fresh Fruit" sample from the SCIEN dataset [61] as acquired by a high-resolution MS imaging device operating in the *Visible Near-Infrared (VNIR)* light range, *i.e.*, sampling light at a wavelength of 400 ~ 900 nm. For this example, the MS cube is resized to $\mathbf{x} \in \mathbb{R}^{n_x \times n_y \times n_\lambda}$, where $n_\lambda = 32$ bands are taken in the *Visible (VIS)* range (400 ~ 700 nm), and the spatial resolution is set to $n_x \times n_y = 128 \times 128$ pixel. This sample is depicted in *Red-Green-Blue (RGB)* colour in Fig. 1.4a and visualised as a data volume in Fig. 1.4b.

Such MS cubes may be decomposed by considering the sparsity basis $\mathbf{D}$ as separable by a Kronecker product $\mathbf{D} = \mathbf{D}_{x,y} \otimes \mathbf{D}_\lambda$ of a spatial-domain basis $\mathbf{D}_{x,y} = \mathbf{D}_x \otimes \mathbf{D}_y$ and a wavelength-domain basis $\mathbf{D}_\lambda$. Provided that all the involved bases are ONBs, $\mathbf{D}$ is unitary and $\mathbf{s} = \mathbf{D}^* \text{vec}(\mathbf{x})$. The optimal choice of these bases will depend on the spatial- and wavelength-domain nature of the scene being considered. While it is reasonable to assume $\mathbf{D}_{x,y}$ as a 2D wavelet basis (*e.g.*, Haar or Daubechies 2D orthonormal wavelets [62]), $\mathbf{D}_\lambda$ depends more specifically on the spectral properties of the MS cube. As an example, some gaseous compounds are known to have line spectra (*e.g.*, the examples in [63]) that may be regarded as sparse in the canonical basis, $\mathbf{D}_\lambda = \mathbf{I}_{n_\lambda}$. Other MS cubes will require a more careful choice of $\mathbf{D}_\lambda$ that yields the sparsest approximation in the wavelength domain, such as scenes that can be sparsely modelled by taking $\mathbf{D}_\lambda$ as the *Discrete Cosine Transform (DCT)* since the corresponding spectral signatures are comprised of smooth profiles.

By choosing $\mathbf{D}_{x,y}$ as the Haar 2D *Discrete Wavelet Transform (DWT)* of $\mathbb{R}^{n_x \times n_y}$, and $\mathbf{D}_\lambda$ as the DCT of $\mathbb{R}^{n_\lambda}$ we obtain the sparse coefficients $\mathbf{s}$ depicted in Fig. 1.5a. While the representation $\mathbf{s}$ of $\mathbf{x}$ w.r.t. $\mathbf{D}$ is not k-sparse, its best k-sparse approximation $\check{\mathbf{s}}$ (Fig. 1.5b) obtained by taking the $k = 19073$ largest magnitude coefficients has an $\text{SNR}_{\check{\mathbf{s}},\mathbf{s}} \approx 40$ dB, *i.e.*, only 4% of $n = 2^{19}$ coefficients represent the MS cube with acceptable dynamic range (Fig. 1.6a); thus, according to Definition 1.4, this signal is $(k = 19073, \vartheta = 0.01)$-compressible.

Alternatively, we may model each *slice* of the MS cube (*i.e.*, the tensor $\mathbf{x}_l \in \mathbb{R}^{n_x \times n_y}$ corresponding to a fixed wavelength index l), mapping $\mathbf{x}$ to $\mathbf{X} = \begin{bmatrix} \text{vec}(\mathbf{x}_0) & \cdots & \text{vec}(\mathbf{x}_{n_\lambda-1}) \end{bmatrix} \in \mathbb{R}^{n_x n_y \times n_\lambda}$. If linear dependencies exist between the n_λ slices $\mathbf{X}$ will be rank-deficient, *i.e.*, $\varrho < n_\lambda$. The low-rank model of Definition 1.8 allows us to exploit the redundancy between slices without requiring any knowledge of the sources (*i.e.*, what spectral signatures combine to form $\mathbf{X}$). For an example, we apply the SVD to $\mathbf{X}$, yielding the singular values reported in Fig. 1.6b. Remarkably, by truncating the decomposition to the $\check{\varrho} = 4$ largest singular values, an approximation $\check{\mathbf{X}}$ having an SNR$_{\check{\mathbf{x}},\mathbf{x}} > 40$ dB is obtained.

Thus, real-world MS cubes are redundant data volumes that comply with the above low-dimensional signal models, and as such may be acquired by sensing operators that verify sparsity-related properties that are introduced in the next Section.

1.2 Sensing Operators and their Properties

As anticipated, $\mathbf{x} \in \mathbb{R}^n$ is here acquired by applying $\mathcal{A}: \mathbb{R}^n \to \mathbb{R}^m, m < n$ that performs a dimensionality reduction which maps $\mathbf{x}$ to $\mathbf{y} = \mathcal{A}(\mathbf{x})$. We here let $\mathcal{A}$ be linear, therefore corresponding to a *sensing matrix* $\mathbf{A} \in \mathbb{R}^{m \times n}$ such that $\mathbf{y} = \mathbf{A}\mathbf{x}$. This *linear encoding* on $\mathbf{x}$ may be implemented in the analog domain, as part of the acquisition of a continuous signal, or in the digital domain, once the samples of $\mathbf{x}$ have already been represented and stored in a quantised form.

Moreover, if $\mathbf{x}$ has a k-sparse representation w.r.t. $\mathbf{D}$ one may write $\mathbf{y} = \mathbf{W}\mathbf{s}$, $\mathbf{W} = \mathbf{A}\mathbf{D} \in \mathbb{R}^{m \times p}$; a recovery algorithm will then be used to revert this mapping and approximate $\mathbf{x}$ by enforcing the sparsity prior.

In this Section we summarise some classical approaches to the design of sensing matrices which rely on the definition of some properties to test whether the mapping of $\mathbf{s} \in \Sigma_k$ to $\mathbf{y}$ occurs without loss of information. These will be connected to essential performance bounds on sparse signal recovery algorithms, emphasising the relationship in Fig. 1.3.

1.2.1 Restricted Isometries

Let us assume for simplicity that $\mathbf{D}$ is an ONB of $\mathbb{R}^n$. Since $\mathbf{W}$ has an $(n-m)$-dimensional kernel (denoted as Ker $(\mathbf{W})$), the Euclidean distance between any two $\mathbf{s}', \mathbf{s}'' \in \mathbb{R}^n$ will not be generally preserved in the corresponding $\mathbf{y}', \mathbf{y}'' \in \mathbb{R}^m$, indicating at least partial information loss. However, when $\mathbf{s}', \mathbf{s}'' \in \Sigma_k$, their distance could still be preserved w.r.t. the union of subspaces in which they lie. In this view, the choice of a sensing operator essentially consists in ensuring that $\mathbf{W}$ is an *approximate isometry* w.r.t. all k-dimensional canonical subspaces of $\mathbb{R}^n$, *i.e.*, that $\mathbf{W}$ must preserve, up to a negligible scaling factor, the distances between k-sparse vectors. This intuition is formalised in the *Restricted Isometry Property (RIP)* [12, 64, 65] as follows.

Definition 1.9 (*Restricted Isometry Property [12]*). Let $\mathbf{W} \in \mathbb{R}^{m \times n}$ be a generic matrix; $\mathbf{W}$ is endowed with the RIP of constant $\delta_k \in [0, 1)$ if

$$\sqrt{1-\delta_k}\|\mathbf{s}\|_2 \leq \|\mathbf{W}\mathbf{s}\|_2 \leq \sqrt{1+\delta_k}\|\mathbf{s}\|_2 \quad (1.8)$$

is verified for all $\mathbf{s} \in \Sigma_k$.

Thus the *Restricted Isometry Constant of order k (k-RIC)*, *i.e.*, δ_k should be as close to 0 as possible; in particular, if any two $\mathbf{s}', \mathbf{s}'' \in \Sigma_k$ of supports T', T'' so that $T' \cap T'' = \emptyset$ are considered, a sufficiently small value of δ_{2k} will ensure that they will be distinguishable by the Euclidean distance $\|\mathbf{y}' - \mathbf{y}''\|_2 > 0$ of their images $\mathbf{y}', \mathbf{y}''$; in a sense, this is analogous to a non-aliasing condition on the images of k-sparse vectors.

In general, finding the value of δ_k amounts to testing, for all k-cardinality supports T, if

$$\sqrt{1-\delta_T}\|\mathbf{s}_T\|_2 \leq \|\mathbf{W}_T\mathbf{s}_T\|_2 \leq \sqrt{1+\delta_T}\|\mathbf{s}_T\|_2 \quad (1.9)$$

holds for some $\delta_T = \max\left\{1 - (\sigma_{\min}(\mathbf{W}_T))^2, (\sigma_{\max}(\mathbf{W}_T))^2 - 1\right\} \in [0, 1)$, *i.e.*, to computing the extreme singular values of $\mathbf{W}_T$ for which

$$\sigma_{\min}(\mathbf{W}_T) \leq \frac{\|\mathbf{W}_T\mathbf{s}_T\|_2}{\|\mathbf{s}_T\|_2} \leq \sigma_{\max}(\mathbf{W}_T) \quad (1.10)$$

This operation is summarised in the following problem.

Problem 1.1 *(Computation of the k-RIC)*. Let $\mathbf{W} \in \mathbb{R}^{m \times n}$ be a generic matrix; its k-RIC is the solution of

$$\delta_k = \max_{T \subseteq \{0,\ldots,n-1\}, |T|=k} \max \left\{ 1 - (\sigma_{\min}(\mathbf{W}_T))^2, (\sigma_{\max}(\mathbf{W}_T))^2 - 1 \right\}$$

As may be expected, testing this property for a generic sensing matrix requires the computation of the singular values of all its $\binom{n}{k}$ k-column submatrices, and is NP-hard [66]. Thus, the verification of the RIP is normally carried out by bounding in probability the extreme singular values in (1.10) when $\mathbf{W}$ is drawn from special *Random Matrix Ensembles (RMEs)* as shown in [17, 65, 67, 68] and recalled in Section 1.2.3. In this view the RIP is only guaranteed in probability, whereas deterministic constructions of $\mathbf{W}$ with this property have been envisioned in [69–72]. Alternative forms and extensions of it also appear in [70, 73, 74].

A number of limits may also be evinced from Definition 1.9, as aside from not being computable the RIP is defined as a *non-adaptive* criterion for k-sparse vectors. In fact (1.9) could hold for $\mathbf{W}$ having generally low $\delta_T \geq \max \left\{ 1 - (\sigma_{\min}(\mathbf{W}_T))^2, (\sigma_{\max}(\mathbf{W}_T))^2 - 1 \right\}$ for typical supports T, with the exception of some critical $\underline{T}$ which force $\delta_k \geq \delta_{\underline{T}}$. Thus, the RIP complies with a *worst-case analysis*, as will be often confirmed by numerical evidence on the effective recoverability of k-sparse vectors. As a consequence RIP-based performance guarantees are usually strict, and this figure of merit is only recommended for its theoretical appeal in connection with the properties of RMEs as will be specified in Section 1.2.3.

1.2.2 Coherence

Another method to assess the information-preserving properties of sensing matrices is defined as follows.

Definition 1.10 *(Coherence [46]).* Let $\mathbf{W} \in \mathbb{R}^{m \times n}$ be a generic matrix of columns $\{\mathbf{w}_j\}_{j=0}^{n-1}$; we define *coherence* (also *mutual coherence* [14]) as

$$\bar{\mu} = \max_{\substack{(j,l) \in \{0,\ldots,n-1\}^2 \\ j \neq l}} \frac{|\mathbf{w}_j^* \mathbf{w}_l|}{\|\mathbf{w}_j\|_2 \|\mathbf{w}_l\|_2}$$

or, when $\overline{\mathbf{W}}$ collects the normalised columns of $\mathbf{W}$,

$$\bar{\mu} = \|\overline{\mathbf{W}}^* \overline{\mathbf{W}} - \mathbf{I}_n\|_\infty \qquad (1.11)$$

Coherence is essentially an index of linear dependence between the columns of $\mathbf{W}$ and should be made as small as possible as to guarantee the recoverability of sparse vectors, as cleared out in Section 1.3.

Moreover, the value of $\bar{\mu}$ can be bounded as follows: since

$$\|\overline{\mathbf{W}}^* \overline{\mathbf{W}}\|_F^2 \geq \frac{\left(\operatorname{tr}\left(\overline{\mathbf{W}}^* \overline{\mathbf{W}}\right)\right)^2}{m} = \frac{n^2}{m}$$

and

$$\|\overline{\mathbf{W}}^* \overline{\mathbf{W}}\|_F^2 \leq n + n(n-1)\bar{\mu}^2$$

we obtain

$$\bar{\mu} \in \left[\sqrt{\frac{n-m}{(n-1)m}}, 1\right]$$

This lower bound was found by Welch [75, 76] and is achieved when $\{\mathbf{w}_j\}_{j=0}^{n-1}$ are so that the off-diagonal entries of $\overline{\mathbf{W}}^* \overline{\mathbf{W}}$ are all identical.

A connection with the RIP may also be made by noting that, using a simple manipulation of (1.8), the k-RIC is also given by

$$\delta_k = \max_{T \subseteq \{0,\ldots,n-1\}, |T|=k} \|\overline{\mathbf{W}}_T^* \overline{\mathbf{W}}_T - \mathbf{I}_k\|_2$$

Since (1.11) may be written equivalently as

$$\bar{\mu} = \max_{T \subseteq \{0,\ldots,n-1\}, |T|=k} \|\overline{\mathbf{W}}_T^* \overline{\mathbf{W}}_T - \mathbf{I}_k\|_\infty$$

and using the inequality between the entrywise ∞-norm and the spectral norm, *i.e.*,

$$\|\overline{\mathbf{W}}_T^* \overline{\mathbf{W}}_T - \mathbf{I}_k\|_\infty \leq \|\overline{\mathbf{W}}_T^* \overline{\mathbf{W}}_T - \mathbf{I}_k\|_2 \leq k \|\overline{\mathbf{W}}_T^* \overline{\mathbf{W}}_T - \mathbf{I}_k\|_\infty$$

it follows that $\delta_k \leq k\bar{\mu}$. Thus, a choice of $(\mathbf{A}, \mathbf{D})$ so that $\bar{\mu}$ achieves its lower bound also hints at a low k-RIC, but the bound on δ_k is not tight[10].

Contrarily to the RIP, coherence has however the benefit of being computable on any instance of $\mathbf{W}$, and is also related to the performances of many signal recovery algorithms [46, 78]; it is therefore recommended as a figure of merit that should be made as small as possible to guarantee that a generic $\mathbf{A}$ and, *e.g.*, an overcomplete dictionary $\mathbf{D}$ allow for the recovery of a sparse vector s.

1.2.3 Sensing Matrices for Compressed Sensing

The properties introduced so far quantify whether $\mathbf{W}$ suitably extracts the information content of a k-sparse s. In particular, both $\mathbf{A}$ and $\mathbf{D}$ contribute to the k-RIC and $\bar{\mu}$; however $\mathbf{A}$ is generally chosen in a form that is convenient at the application level, whereas $\mathbf{D}$ is set depending on the sparsity of x w.r.t. it. To summarize, it is desirable to choose $\mathbf{A}$ independently of $\mathbf{D}$ and so that $\mathbf{W}$ is still endowed with either low coherence or the RIP; to do so, rather than recurring to special deterministic designs $\mathbf{W}$ is generally drawn as a random matrix from one of the RMEs described below.

Random Matrix Ensembles

We here list the most common RMEs known to verify the aforementioned properties (see [17, 65, 79]). A normalisation factor (*e.g.*, $1/\sqrt{m}$ or $1/\sqrt{n}$) on $\mathbf{W}$ may also be considered.

- *Isotropic Random Gaussian Ensemble (RGE):* $\mathbf{W} \in \mathbb{R}^{m \times n}$ with *independent and identically distributed (i.i.d.)* entries following a univariate Gaussian distribution, *i.e.*, $\forall (j, l) \in \{0, \ldots, m-1\} \times \{0, \ldots, n-1\}$, $W_{j,l} \sim \mathcal{N}\left(0, \frac{1}{m}\right)$;

- *Isotropic Random Bernoulli Ensemble (RBE):* $\mathbf{W} \in \{-1, +1\}^{m \times n}$ with i.i.d. entries following a symmetric Bernoulli distribution,

---

[10]A slightly sharper result may be found by using Gershgorin's circle theorem [77] as indicated in [72].

i.e., $\forall(j,l) \in \{0,\ldots,m-1\} \times \{0,\ldots,n-1\}$, $\mathbb{P}[W_{j,l} = -1] = \mathbb{P}[W_{j,l} = +1] = \frac{1}{2}$;

- *Random sub-Gaussian Ensemble (RsGE)*: $\mathbf{W} \in \mathbb{R}^{m \times n}$ with either (i) i.i.d. entries or (ii) i.i.d. row vectors following a *sub-Gaussian* distribution, *i.e.*, whose moments are bounded by those of a suitable Gaussian distribution. One such example, aside from the RGE and RBE, is the *Isotropic Random Ternary Ensemble (RTE)*, *i.e.*, $\mathbf{W} \in \{-1,0,1\}^{m \times n}$ with i.i.d. entries distributed as $\forall(j,l) \in \{0,\ldots,m-1\} \times \{0,\ldots,n-1\}, \mathbb{P}[W_{j,l} = -1] = \mathbb{P}[W_{j,l} = +1] = \frac{1}{6}, \mathbb{P}[W_{j,l} = 0] = \frac{2}{3}$;

- *Partial Fourier Ensemble (PFE)*: $\mathbf{W} \in \mathbb{C}^{m \times n}$ whose rows $\{\mathbf{w}_j\}_{j=0}^{m-1}$ are chosen uniformly at random from the basis vectors of the DFT matrix $\mathbf{F}_n$ of entries $F_{j,l} = e^{-\iota \frac{2\pi}{n} jl}$, *i.e.*, $\mathbf{W} = \mathbf{P}^{\Omega} \mathbf{F}_n$ with $\mathbf{P}^{\Omega}$ a *selection matrix*[11];

- *Partial Hadamard Ensemble (PHE)*: $\mathbf{W} \in \{-1,+1\}^{m \times n}$ whose rows $\{\mathbf{w}_j\}_{j=0}^{m-1}$ are chosen uniformly at random from the basis vectors of the Hadamard matrix[12] $\mathbf{H}_n$, *i.e.*, $\mathbf{W} = \mathbf{P}^{\Omega} \mathbf{H}_n$.

Clearly $\mathbf{W}$ will have different k-RIC values depending on the chosen RME, with the RGE generally being taken as a reference for achieving minimum δ_k for fixed (m,n). In addition, the sensing matrix applied on $\mathbf{x}$ (*i.e.*, in the signal domain) is $\mathbf{A}$, so for $\mathbf{W}$ to belong to the above ensembles the pair $(\mathbf{A}, \mathbf{D})$ must be carefully chosen: in the case of a PHE (or PFE) either $\mathbf{A}$ is drawn from the ensemble and $\mathbf{D} = \mathbf{I}_n$, vice versa $\mathbf{A} = \mathbf{P}^{\Omega}$ and $\mathbf{D} = \mathbf{H}_n$ (or $\mathbf{F}_n$); in most other cases $\mathbf{A}$ will be drawn from the above ensembles, and it must be verified that the rotation of the rows of $\mathbf{A}$ caused by $\mathbf{D}$ does not significantly increase δ_k or $\bar{\mu}$. It is also worth noting that the RIP can be slightly modified to include the case of overcomplete $\mathbf{D}$ (see [50, 80]).

---

[11] An $m \times n$ selection matrix $\mathbf{P}^{\Omega}$ is obtained by extracting m rows corresponding to the random indices in $\Omega \subseteq \{0,\ldots,n-1\}, |\Omega| = m$ from the identity matrix $\mathbf{I}_n$.

[12] Hadamard matrices of order n, *i.e.*, $\mathbf{H}_n \in \{-1,+1\}^{n \times n}$ so that $\mathbf{H}_n^* \mathbf{H}_n = n\mathbf{I}_n$, were shown to exist for $n = 4q, q \in \mathbb{Z}_+$. The $\mathcal{O}(n \log n)$ complexity computation of this transform requires $n = 2^q, q \in \mathbb{Z}_+$.

Finally, it must be noted that many other RMEs were shown to have the above properties: among a growing literature, a proof of the RIP for circulant random matrices related to convolution operators appears in [31, 81].

Sub-Gaussian Random Matrix Ensembles and the RIP

A common case in which RMEs are endowed with the RIP is that of RsGEs: the purpose of this Section is to justify this statement with a sufficient level of detail. In particular, we now specify *in a simplified fashion* the elements that form a RsGE.

Definition 1.11 *(Sub-Gaussian random variable).* Let $w \in \mathbb{R}$ be a *Random Variable (r.v.)* with[a] *Probability Density Function (PDF)* $f(w)$ and $\mu_w = 0$, and recall the *Moment-Generating function (MGF)* of a r.v. $g \sim \mathcal{N}(0, \sigma^2)$, $\mathbb{E}[e^{\tau g}] = e^{\frac{\tau^2 \sigma^2}{2}}, \tau \in \mathbb{R}$; w $(f(w))$ is a *sub-Gaussian r.v.* (distribution) if there exists a finite quantity $\sigma \geq 0$ such that $\forall \tau \in \mathbb{R}, \mathbb{E}[e^{\tau w}] \leq e^{\tau^2 \sigma^2}$.

This definition is extended to *Random Vectors (R.V.s)* as follows.

Definition 1.12 *(Sub-Gaussian Random Vector).* Let $\mathbf{w} \in \mathbb{R}^n$ be a R.V. with PDF $f(\mathbf{w})$; $\mathbf{w}$ is *sub-Gaussian* if for all $T \subseteq \{0, \ldots, n-1\}$ the marginal PDFs $f_{\mathbf{w}_T}(\mathbf{w}_T)$ are sub-Gaussian.

Noteworthy distributions agreeing with this definition are (i) multivariate Gaussian R.V.s, *i.e.*, $\mathbf{w} \sim \mathcal{N}(\boldsymbol{\mu}_\mathbf{w}, \mathbf{K}_\mathbf{w})$ and (ii) symmetric Bernoulli ones, *i.e.*, $\mathbf{w} \in \{-1, +1\}^n : \forall j \in \{0, \ldots, n-1\}, \mathbb{P}[w_j = -1] = \mathbb{P}[w_j = +1] = \frac{1}{2}$. Thus, constructing a matrix $\mathbf{W}$ (or $\mathbf{A}$) as a collection of sub-Gaussian entries or row vectors forms a RsGE.

Another important property of sub-Gaussian R.V.s is that they are *rotationally invariant* [82, Lemma 5.9], *i.e.*, if a R.V. $\mathbf{a} \in \mathbb{R}^n$ is sub-Gaussian then for any $\mathbf{d} \in \mathbb{R}^n \setminus \{\mathbf{0}_n\}$, $w = \mathbf{a}^*\mathbf{d}$ is in turn a sub-Gaussian r.v. (in analogy with the rotational invariance of the multivariate Gaussian distribution). This property is crucial since we have so far referred to $\mathbf{W}$ drawn from a RsGE, while $\mathbf{A}$ will truly be generated by it in practice. Due to rotational invariance, $\forall j \in \{0, \ldots, m-1\}, l \in \{0, \ldots, n-1\}, W_{j,l} = \mathbf{a}_j^* \mathbf{d}_l$ will still have a sub-Gaussian distribution, so we may claim that if $\mathbf{D}$ is an ONB $\mathbf{W} = \mathbf{A}\mathbf{D}$ is a RsGE.

The relevance of RsGEs to verifying the RIP follows from the next statements, which relate the extreme singular values appearing in (1.10) with specific RsGEs.

Theorem 1.3 (*Singular Values of RsGEs with i.i.d. rows [82, Theorem 5.39]*). Let $\mathbf{W} \in \mathbb{R}^{m \times k}$ be a RsGE whose rows $\{\mathbf{w}_j\}_{j=0}^{m-1}$ are i.i.d. copies of a sub-Gaussian R.V. $\mathbf{w} \in \mathbb{R}^k$ with $\mathbf{K_w} = \mathbf{I}_k$. For any $\tau \geq 0$,

$$\mathbb{P}\left[\sigma_{\min}(\mathbf{W}) \geq \sqrt{m} - c_w\sqrt{k} - \tau\right] \geq 1 - 2e^{-c'_w \tau^2}$$
$$\mathbb{P}\left[\sigma_{\max}(\mathbf{W}) \leq \sqrt{m} + c_w\sqrt{k} + \tau\right] \geq 1 - 2e^{-c'_w \tau^2} \quad (1.12)$$

with c_w, c'_w positive constants depending on the MGF of $\{\mathbf{w}_j\}_{j=0}^{m-1}$.

In particular, this improves on previous asymptotic results [83–85].

Theorem 1.4 (*Singular Values of RMEs with i.i.d. entries [85, Theorem 1]*). Let $\mathbf{W} \in \mathbb{R}^{m \times k}$ be a RME whose i.i.d. r.v.s $W_{j,l}$ are such that $\forall (j,l) \in \{0, \ldots, m-1\} \times \{0, \ldots, k-1\}, \mu_{W_{j,l}} = 0, \sigma^2_{W_{j,l}} = 1, \mathbb{E}[W_{j,l}^4] < +\infty$. For $m, k \to \infty$ as $k/m \to r$,

$$\mathbb{P}[\sqrt{m} - \sqrt{k} \leq \sigma_{\min}(\mathbf{W}) \leq \sigma_{\max}(\mathbf{W}) \leq \sqrt{m} + \sqrt{k}] \simeq 1 \quad (1.13)$$

A numerical evaluation of these singular values' distribution is reported for the RGE (Fig. 1.7) and RBE (Fig. 1.8) with $m = 2^8, k = \{2, 4, \ldots, 128, 192\}$ and over 2^{20} trials. As can be seen from Fig. 1.7b, 1.8b the extreme singular values detach from 1 for both RMEs as k approaches m (i.e., the k-RIC corresponding to a submatrix of such dimensions increases).

---

[a]This definition can be extended to variables with $\mu_w \neq 0$, however only the zero-mean case is here reviewed for the sake of simplicity.

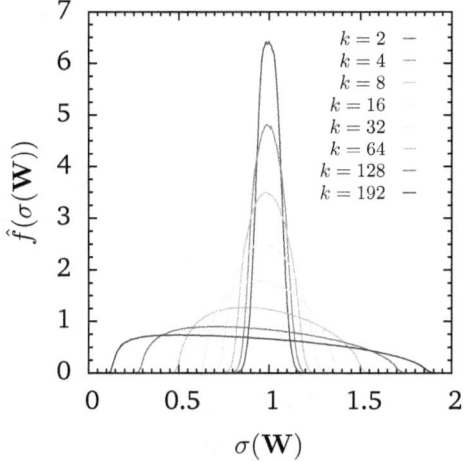

(a) Empirical PDF of singular values; asymptotically, it follows the Marchenko-Pastur law [86].

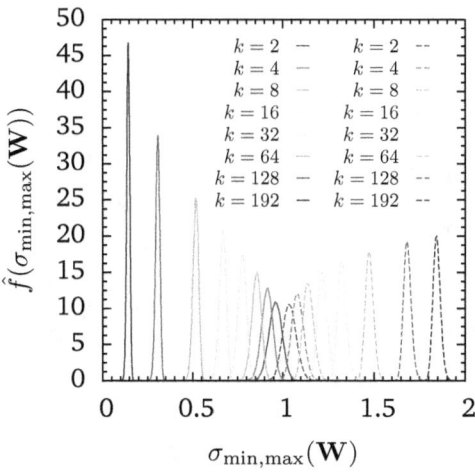

(b) Extreme singular values' PDF: $\sigma_{\min}(\mathbf{W})$ (solid), $\sigma_{\max}(\mathbf{W})$ (dashed).

Figure 1.7: Empirical distribution of the singular values of the RGE corresponding to a k-column submatrix, with $m = 2^8$ as k varies.

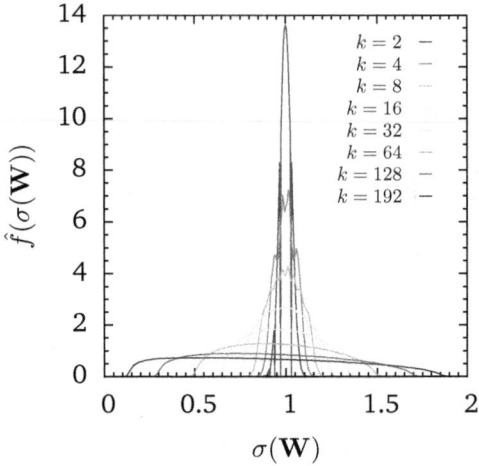

(a) Empirical PDF of singular values; asymptotically, it follows the Marchenko-Pastur law [86].

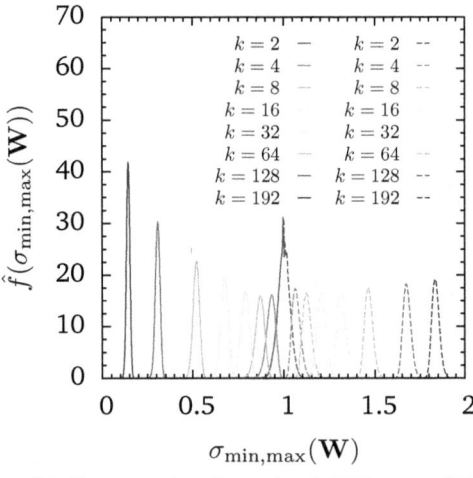

(b) Extreme singular values' PDF: $\sigma_{\min}(\mathbf{W})$ (solid), $\sigma_{\max}(\mathbf{W})$ (dashed).

Figure 1.8: Empirical distribution of the singular values of the RBE corresponding to a k-column submatrix, with $m = 2^8$ as k varies.

By comparing (1.12) (or (1.13)) with (1.9) a strategy to verifying the RIP in probability emerges as follows. Since Theorem 1.3 is applied to all k-column submatrices $\mathbf{W}_T \in \mathbb{R}^{m \times k}$ of a RsGE $\mathbf{W} \in \mathbb{R}^{m \times n}$, Thus $\mathbf{W}_T$ is plugged in (1.12), and by taking the union bound over all T the RIP can be bounded in probability. Omitting the technical details of this proof (found in [82]) the following result concludes this review of sub-Gaussian matrix ensembles and their importance for CS.

Theorem 1.5 (*RIP of the RsGE [82]*). Let $\mathbf{W} \in \mathbb{R}^{m \times n}$ be a RsGE whose rows $\{\mathbf{w}_j\}_{j=0}^{m-1}$ are i.i.d. copies of a sub-Gaussian R.V. of $\mathbb{R}^n$ with $\mathbf{K}_{\mathbf{w}_j} = \frac{1}{\sqrt{m}}\mathbf{I}_n$. Then for any $k \in \{1, \ldots, n\}$ and a value $\delta \in (0,1)$

$$m \geq \overline{m} \Rightarrow \mathbb{P}[\delta_k \leq \delta] \geq 1 - 2e^{-c_w m \delta^2}$$
$$\overline{m} = c'_w \frac{k}{\delta^2} \log\left(e\frac{n}{k}\right) \tag{1.14}$$

with c_w, c'_w positive constants depending on the MGF of $\{\mathbf{w}_j\}_{j=0}^{m-1}$.

Thus, once the above constants are found and arbitrary values (δ, k) are set, it is possible to find a critical value $\overline{m} = \mathcal{O}(k \log n/k)$ (*i.e.*, a minimum number of measurements) for which the RIP is guaranteed with $\delta_k \leq \delta$ and high probability. Moreover, once $\mathbf{A}$ is drawn from a RsGE to have the RIP, by rotational invariance $\mathbf{W}$ maintains the same k-RIC for any ONB $\mathbf{D}$ [65]. Since this choice of $\mathbf{A}$ is agnostic of $\mathbf{D}$, RsGEs are said to provide a *universal* encoding [17].

Other Ensembles and the RIP

Clearly, the PFE and PHE as mentioned above are not RsGEs (*i.e.*, attempting to qualify them as such leads to very weak moment bounds). For these specific RMEs it can be shown by a similar procedure [82, Theorem 5.71] that $m \geq \overline{m} \Rightarrow \mathbb{E}[\delta_k] \leq \delta$ for a minimum of

$$\overline{m} = \mathcal{O}\left(\frac{k}{\delta^2} \log(n) \log^3(k)\right) \tag{1.15}$$

randomly selected DFT or Hadamard measurements. By qualitative comparison of (1.14) with (1.15) it is clear that RsGE matrices are capable of providing the RIP with relatively smaller undersampling rates m/n (although the value of c'_w should be validated for finite dimensions); nevertheless, the low implementation complexity of the PHE and the PFE will make them appealing alternatives in many applications.

As a remark, to extend and improve their applicability so that $\mathbf{A}$ can be drawn as a PFE or PHE and $\mathbf{D}$ may be arbitrary, an additional randomisation step may be introduced to produce two new RMEs by means of a diagonal matrix $\mathbf{B} \in \{-1, 0, 1\}^{n \times n}$ of i.i.d. non-zero entries $\forall i \in \{0, \ldots, n-1\}, \mathbb{P}[B_{i,i} = +1] = \mathbb{P}[B_{i,i} = -1] = \frac{1}{2}$, i.e., $\mathbf{A} = \mathbf{P}^\Omega \mathbf{H}_n \mathbf{B}$ (similarly to [87, 88]) and

$$\mathbf{A} = \mathbf{P}^\Omega \mathbf{F}_n \mathbf{B} \qquad (1.16)$$

which will be used as a computationally efficient reference in Chapter 9 (this is also known as a spread-spectrum ensemble in [89]).

Summarising, we conclude that the reviewed ensembles show a common *concentration* feature that makes (1.8) hold for most $\mathbf{s} \in \mathbb{R}^n$ with a minimum of $\overline{m} = \mathcal{O}(k \log n)$ measurements. This avoids the verification of δ_k by Problem 1.1, yet some limits of the RIP approach stand, as (1.14) and (1.15) depend on some universal constants which should be anyway numerically verified for finite (m, n) by a large-scale Monte Carlo simulation. Thus, we will ultimately rely on a numerical evaluation to find $\overline{m}$ for the chosen setting based on the actual capability of recovering $\mathbf{s}$ from $\mathbf{y} = \mathbf{W}\mathbf{s}$. This empirical approach is supported by a peculiar geometric behaviour discussed in the next Section.

1.2.4 Donoho-Tanner Phase Transition

While the properties of Definition 1.9, 1.10 have a simple algebraic form, the *Donoho-Tanner phase transition* [18, 79, 90] requires deeper mathematical notions. In words, this phase transition is observed when assessing the solvability of a sparsity-promoting convex optimisation problem as its dimensions vary. This problem is here kept implicit

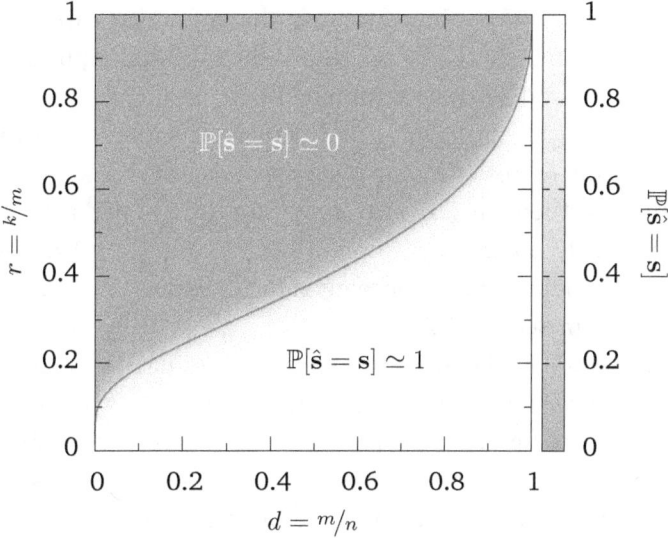

Figure 1.9: The phase space (d, r); the Donoho-Tanner phase transition curve at $\mathbb{P}[\hat{\mathbf{s}} = \mathbf{s}] = \frac{1}{2}$ for RGE matrices is overlaid in red.

and denoted as a function $\Delta_{\mathbf{y}=\mathbf{Ws}}$ yielding an estimate $\hat{\mathbf{s}}$ that is a counterimage of the measurements $\mathbf{y}$ w.r.t. $\mathbf{W}$.

In more detail, we let $\mathbf{W}$ be drawn from a RME and $\mathbf{s}$ be any sparse vector, then consider a diagram (*phase space*, see Fig. 1.9) reporting the probability that any instance of $\mathbf{y} = \mathbf{Ws}$ is solved by $\hat{\mathbf{s}} = \Delta_{\mathbf{y}=\mathbf{Ws}}(d, r)$ as a function of $d = m/n \in (0, 1)$ and $r = k/m \in (0, 1)$, *i.e.*, that the sparse solution to any such underdetermined linear system of equations is found. A sharp transition curve is empirically observed [79] on the phase space when evaluating $\mathbb{P}[\hat{\mathbf{s}} = \mathbf{s}]$: this curve separates a "0" region (*i.e.*, $\mathbb{P}[\hat{\mathbf{s}} = \mathbf{s}] \simeq 0$) from a "1" region (*i.e.*, $\mathbb{P}[\hat{\mathbf{s}} = \mathbf{s}] \simeq 1$), its shape depending on the objective of $\Delta_{\mathbf{y}=\mathbf{Ws}}$ as well as the RME from which $\mathbf{W}$ is drawn.

Remarkably, this curve is connected with the number of k-dimensional faces of a convex polytope $Q \subseteq \mathbb{R}^n$ that survive[13] after

---

[13] Projection by means of $\mathbf{W}$ may only reduce the number of faces.

random projection by $\mathbf{W}$ as k-dimensional faces of $P = \{\mathbf{y} \in \mathbb{R}^m : \mathbf{y} = \mathbf{W}\mathbf{s}, \mathbf{s} \in Q\}$. In particular, the reference case here takes $Q = \{\mathbf{x} \in \mathbb{R}^n : \|\mathbf{x}\|_1 \leq 1\}$ (the n-dimensional equivalent of an octahedron) and $\mathbf{W}$ drawn from the RGE.

Asymptotically in (m, n, k) and as (d, r) vary, the ratio between the k-dimensional faces of P and Q corresponds to $\mathbb{P}[\hat{\mathbf{s}} = \mathbf{s}], \|\mathbf{s}\|_0 = k$ [79, Lemma 2.2, Theorem 2.3]. Even more surprisingly, the same behaviour is empirically observed for finite n at $\mathbb{P}[\hat{\mathbf{s}} = \mathbf{s}] = \frac{1}{2}$ [90] and does not change substantially for many non-Gaussian RMEs such as those mentioned in Section 1.2.3.

Aside from the theoretical depth of these results (only outlined here for the sake of brevity) the practical relevance of this approach is in that it provides the tightest estimates w.r.t. the recoverability of a k-sparse s as the dimensions of $\mathbf{W}$ vary; this is highlighted in [90, Section X]. In fact, when compared with the Donoho-Tanner phase transition, the RIP only provides a lower bound, *i.e.*, it delineates a significantly smaller "1" region in the phase space [73]. An empirical evaluation of the phase transition of $\Delta_{\mathbf{y}=\mathbf{W}\mathbf{s}}(d, r)$ w.r.t. $\mathbf{W}$ from different RMEs is in many senses preferable to attempting a computation of the RIP. The clear drawback of this approach is that it still requires a large computational effort (*i.e.*, a large-scale Monte Carlo simulation) to accurately trace the phase space, and it must be repeated for any finite n.

1.3 Signal Recovery Algorithms

The purpose of the properties introduced in Section 1.2 is to find $\mathbf{A}$ ($\mathbf{W}$) so that the linear encoding $\mathbf{y} = \mathbf{A}\mathbf{x}$ ($\mathbf{y} = \mathbf{W}\mathbf{s}$) robustly embeds the information in $\mathbf{x}$; in order to revert it, some optimisation problems are here solved by enforcing one of the signal models in Section 1.1, yielding an estimate $\hat{\mathbf{x}}$ (or equivalently, of its sparse representation $\hat{\mathbf{s}}$). In particular, the models in Definition 1.3-1.5,1.7,1.8 imply non-convex objective functions. Broadly speaking, convexity is fundamental in the solution of mathematical programming problems, as it guarantees the existence of a global minimum and of numerical methods to achieve it exactly. Thus, the general strategy adopted to promote such models is *convex relaxation, i.e.,* their non-convex objective functions are relaxed into convex ones. A particular emphasis is deserved by the relaxation of $\|\cdot\|_0$, which measures sparsity in Definition 1.3, to $\|\cdot\|_1$. Fundamental results exist [12, 16] implying the so-called "$\ell_0 = \ell_1$" equivalence, *i.e.,* the fact that the k-sparse solution to an underdetermined linear system can be found by means of this relaxation. Remarkably, the conditions for this equivalence may be verified in terms of any of the properties in Section 1.2.

These problems shall also take into account the corruption caused by additive noise $\nu \in \mathbb{R}^m$ on the measurements, *i.e.,* $\mathbf{y} = \mathbf{Ax} + \nu$, noting that this noise term may also be dependent w.r.t. x. Such disturbances are inevitable when a physical or finite-precision implementation of a sensing operator is considered. Even in this case, guarantees exist [13, 64, 91] and ensure that at least a bounded-error approximation $\hat{s}$ of s can be recovered when $\mathbf{y}$ is acquired by a suitable sensing matrix.

In terms of algorithms to solve such convex problems, different approaches have been used in this thesis and are explicitly specified where needed; as a general guideline, once a convex formulation of a problem is found (eventually testing it with the aid of modelling languages such as CVX [92]), frameworks such as GUROBI [93] or CPLEX [94] allow its solution by means of general solvers that handle linear or quadratic programming (see [95]). In particular, a variety of solvers is available for problems in continuous, integer or mixed variables, linear or quadratic objective functions and whose solution space is a convex set. Although the computational complexity of such problems is polynomial in their dimensionality, the general solvers' computational requirements as n increases suggest that specific algorithms should be adopted for very large-scale optimisation. In this context, solvers such as *proximal algorithms* (see [96, 97]) or similarly *projected gradient-based methods* [98] have gained popularity for their efficiency and robustness.

Nevertheless, different perspectives on signal recovery exist, the most popular alternatives to convex problems being either *greedy optimisation* techniques, *i.e.*, iterative algorithms that promote the above models by computing a local heuristic (*i.e.*, a proxy) at each iteration [99–102] as well as *probabilistic inference algorithms* [22, 103–105]. Since our purpose is to adapt the design of sensing matrices to specific tasks, only the principles, problems and guarantees of sparse signal recovery by convex optimisation are reviewed below with the objective of providing a complete background for a fair evaluation of special matrix designs.

1.3.1 Sparse Signal Recovery by Convex Optimisation

Sparse Signal Recovery

We now recall some forms of the signal recovery problem that depend on the amount of information available to the *decoder*, *i.e.*, the algorithm that processes $\mathbf{y}$ to yield $\hat{\mathbf{s}}$. Throughout this section, we let $\mathbf{x} \in \mathbb{R}^n, \mathbf{s} \in \mathbb{R}^p : \mathbf{x} = \mathbf{Ds}, \mathbf{D} \in \mathbb{R}^{n \times p}; \mathbf{y} \in \mathbb{R}^m : \mathbf{y} = \mathbf{Ax}, \mathbf{A} \in \mathbb{R}^{m \times n}$ or equivalently $\mathbf{y} = \mathbf{Ws}, \mathbf{W} \in \mathbb{R}^{m \times p}$. A common approach to approximating the solution of a non-invertible system of equations is that of *linear least-squares*, yielding the minimum 2-norm solution by means of the Moore-Penrose pseudoinverse.

Problem 1.2 (*Linear Least-Squares*). The *Linear Least-Squares (LLS)* solution to an underdetermined linear system of equations $\mathbf{y} = \mathbf{Ws}$ is

$$\hat{\mathbf{s}} = \operatorname*{argmin}_{\boldsymbol{\xi} \in \mathbb{R}^n} \|\boldsymbol{\xi}\|_2 \text{ s.t. } \mathbf{y} = \mathbf{W}\boldsymbol{\xi} \tag{1.17}$$

whose closed-form expression $\hat{\mathbf{s}} = \mathbf{W}^\dagger \mathbf{y}$.

An interpretation of this problem in $\mathbb{R}^3$ is reported in Fig. 1.10a; from this, it can be concluded that adopting $\|\cdot\|_2$ as the objective function tends to promote the unique *minimum energy* solution regardless of the sparsity of $\mathbf{s}$. Clearly, this does not contemplate the fact that the exact $\mathbf{s}$ is sparse, and $\|\hat{\mathbf{s}} - \mathbf{s}\|_2$ may become arbitrarily large. Thus, this approach is clearly unsuitable for sparse signal recovery.

Moving to a formulation that leverages sparsity, once it is known that a k-sparse solution exists (*i.e.*, that $k < m$ and $\operatorname{rank}(\mathbf{W}) = m$), its unicity must also be verified.

Theorem 1.6 *(Uniqueness of a k-sparse solution [15, 106])*. Let $\mathbf{y} = \mathbf{Ws}$ have a solution $\mathbf{s}$, $\|\mathbf{s}\|_0 = k$ and let $\bar{\mu}$ be the coherence of $\mathbf{W}$. If $k < \frac{1}{2}\left(1 + \frac{1}{\bar{\mu}}\right)$ then $\forall \mathbf{s}' \in \mathbb{R}^n$, $\mathbf{s}' \neq \mathbf{s} : \mathbf{y} = \mathbf{Ws}'$, $\|\mathbf{s}'\|_0 > k$.

When such a sparse solution is sought under these basic conditions, we face the following combinatorial problem.

Problem 1.3 *(Sparsest Solution of a Linear System)*. The sparse solution to an underdetermined linear system of equations $\mathbf{y} = \mathbf{Ws}$ is

$$\mathbf{s} = \Delta_0(\mathbf{y}, \mathbf{W}) = \underset{\boldsymbol{\xi} \in \mathbb{R}^n}{\operatorname{argmin}} \|\boldsymbol{\xi}\|_0 \text{ s.t. } \mathbf{y} = \mathbf{W}\boldsymbol{\xi} \qquad (1.18)$$

Problem 1.3 is clearly equivalent to testing whether

$$\boldsymbol{\xi} = \begin{cases} \boldsymbol{\xi}_T = (\mathbf{W}_T)^\dagger \mathbf{y}, & T = \operatorname{supp}(\boldsymbol{\xi}), |T| = k \\ \boldsymbol{\xi}_{T^c} = \mathbf{0}_{n-k}, & T^c = \{0, \ldots, n-1\} \setminus \operatorname{supp}(\boldsymbol{\xi}) \end{cases}$$

is such that $\mathbf{y} = \mathbf{W}\boldsymbol{\xi}$ for each sparsity level k and over $\binom{n}{k}$ possible supports T. This problem is also known as ℓ_0-*minimisation*, and is NP-hard (it is identical to (1.3)). If, however, an *oracle* supplied $\overline{T} = \operatorname{supp}(\mathbf{s})$ the problem would be substantially solved.

Problem 1.4 *(Oracle Least-Squares)*. The *Oracle Least-Squares (OLS)* solution to an underdetermined linear system of equations $\mathbf{y} = \mathbf{Ws}$ where $\mathbf{s} \in \Sigma_k$ and $\overline{T} = \operatorname{supp}(\mathbf{s})$ is known *a priori* is

$$\hat{\mathbf{s}} = \Delta_{\operatorname{OLS}}(\mathbf{y}, \mathbf{W}, \overline{T}) = \underset{\boldsymbol{\xi} \in U_{\overline{T}}}{\operatorname{argmin}} \|\boldsymbol{\xi}\|_2 \text{ s.t. } \mathbf{y} = \mathbf{W}\boldsymbol{\xi} \qquad (1.19)$$

whose closed-form expression

$$\hat{\mathbf{s}} = \begin{cases} \hat{\mathbf{s}}_{\overline{T}} = (\mathbf{W}_{\overline{T}})^{\dagger}\mathbf{y}, & \overline{T} = \mathrm{supp}(\mathbf{s}) \\ \hat{\mathbf{s}}_{\overline{T}^c} = \mathbf{0}_{n-k}, & \overline{T}^c = \{0, \ldots, n-1\} \setminus \mathrm{supp}(\mathbf{s}) \end{cases} \quad (1.20)$$

The simple geometry of this problem is reported in Fig. 1.10b: once the subspace U_T in which s lies is known, it can be trivially recovered as in (1.20). Thus, the problem of inferring it from $\mathbf{y}$ is only a matter of knowing $\overline{T} = \mathrm{supp}(\mathbf{s})$; many algorithms leverage local heuristics that aim to detect $\overline{T}$ [46, 99, 100, 107, 108].

We now perform a convex relaxation of $\|\cdot\|_0$ in (1.18) by replacing it with $\|\cdot\|_1$ (*i.e.*, the closest convex p-norm); in this case, the solution of the relaxed problem will be an approximation of s, eventually equalling it under special conditions on the geometry of the underdetermined system. This relaxed, ℓ_1-*minimisation* problem follows.

Problem 1.5 (*Basis Pursuit*). The *Basis Pursuit (BP)* solution to an underdetermined linear system of equations $\mathbf{y} = \mathbf{W}\mathbf{s}$ is

$$\hat{\mathbf{s}} = \Delta_{\mathrm{BP}}(\mathbf{y}, \mathbf{W}) = \underset{\boldsymbol{\xi} \in \mathbb{R}^n}{\mathrm{argmin}} \, \|\boldsymbol{\xi}\|_1 \text{ s.t. } \mathbf{y} = \mathbf{W}\boldsymbol{\xi} \quad (1.21)$$

The analysis form of this problem is

$$\hat{\mathbf{x}} = \underset{\boldsymbol{\xi} \in \mathbb{R}^n}{\mathrm{argmin}} \, \|\mathbf{D}^{\star}\boldsymbol{\xi}\|_1 \text{ s.t. } \mathbf{y} = \mathbf{A}\boldsymbol{\xi} \quad (1.22)$$

with $\mathbf{D}^{\star} \in \mathbb{R}^{n \times p}$ the corresponding analysis transform.

A geometric intuition of this problem is reported in Fig. 1.11a; it is now clear that the convex polytope Q anticipated in Section 1.2.4 is the *cross-polytope* of $\mathbb{R}^n$ that delineates the level curves of the objective function in (1.21). It may now be specified that the Donoho-Tanner phase transition is indeed observed for this convex problem, and that the values of $\mathbb{P}[\mathbf{s} = \Delta_{\mathrm{BP}}(\mathbf{y}, \mathbf{W})]$ are those that guarantee, at least in probability, the equivalence $\hat{\mathbf{s}} = \Delta_{\mathrm{BP}}(\mathbf{y}, \mathbf{W}) = \Delta_0(\mathbf{y}, \mathbf{W}) = \mathbf{s}$. In particular, the procedure of evaluating the empirical phase transition of

BP when $\mathbf{W}$ is varied for different (m, n, k) will subtend some findings in this thesis.

Moreover, the analysis form of BP may also be adopted when overcomplete dictionaries $\mathbf{D}$ are considered. In this case the geometry of the problem is slightly different, as $\|\cdot\|_1$ is not enforced in the signal domain but in that of $\mathbf{D}^\star$. TV minimisation also falls in this case, replacing $\|\cdot\|_1$ by $\|\cdot\|_{\text{TV}}$ (evaluated in a suitable number of tensor dimensions).

The process of acquiring measurements $\mathbf{y} = \mathbf{A}\mathbf{x} + \boldsymbol{\nu}$ about some quantity $\mathbf{x}$ is always, to some extent, subject to noise sources here summarised as $\boldsymbol{\nu} \in \mathbb{R}^m$. Such disturbances may be due to physical-level signal-independent phenomena such as thermal effects in the acquisition device, as well as signal-dependent ones such as quantisation, multiplicative noise or miscalibration effects on $\mathbf{A}$. For this reason the equality condition $\mathbf{y} = \mathbf{A}\mathbf{x}$ is often relaxed to a *data fidelity* constraint, $\|\mathbf{y} - \mathbf{A}\mathbf{x}\|_2 \leq \varepsilon$, admitting some uncertainty between the measurements and the actual solution. This translates into the following optimisation problem.

Problem 1.6 *(Basis Pursuit with Denoising)*. The *Basis Pursuit with Denoising (BPDN)* solution to an underdetermined linear system of equations $\mathbf{y} = \mathbf{W}\mathbf{s} + \boldsymbol{\nu}$ affected by additive noise such that $\|\boldsymbol{\nu}\|_2 \leq \varepsilon$ is

$$\hat{\mathbf{s}} = \Delta_{\text{BPDN}}(\mathbf{y}, \mathbf{W}, \varepsilon) = \underset{\boldsymbol{\xi} \in \mathbb{R}^n}{\arg\min} \|\boldsymbol{\xi}\|_1 \text{ s.t. } \|\mathbf{y} - \mathbf{W}\boldsymbol{\xi}\|_2 \leq \varepsilon \quad (1.23)$$

The analysis form of this problem is

$$\hat{\mathbf{x}} = \underset{\boldsymbol{\xi} \in \mathbb{R}^n}{\arg\min} \|\mathbf{D}^\star \boldsymbol{\xi}\|_1 \text{ s.t. } \|\mathbf{y} - \mathbf{A}\boldsymbol{\xi}\|_2 \leq \varepsilon \quad (1.24)$$

with $\mathbf{D}^\star \in \mathbb{R}^{n \times p}$ the corresponding analysis transform.

A geometric intuition of this problem is reported in Fig. 1.11b. There, the quadratic constraint replaces the flat $\mathbf{y} = \mathbf{W}\mathbf{s}$ by an n-dimensional tube of radius ε, $\mathcal{T}_\varepsilon$. Clearly, the *noise threshold* ε should be suitably chosen for each problem instance to exceed $\|\boldsymbol{\nu}\|_2$ so that $\mathbf{s}$ is still in the solution space of Problem 1.6. Estimating ε is therefore important: a *"genie"-tuning* will be mentioned in this thesis, *i.e.*, when $\varepsilon^\star = \|\boldsymbol{\nu}\|_2$ is known exactly for each problem instance.

A further relaxation of BPDN (that in fact preceded it as a method for sparse statistical regression [109]) is given by the following problem.

Problem 1.7 *(Least Absolute Shrinkage and Selection Operator)*. The *Least Absolute Shrinkage and Selection Operator (LASSO)* solution to an underdetermined linear system of equations $\mathbf{y} = \mathbf{W}\mathbf{s}$ or $\mathbf{y} = \mathbf{W}\mathbf{s} + \boldsymbol{\nu}$ is

$$\hat{\mathbf{s}} = \underset{\boldsymbol{\xi} \in \mathbb{R}^n}{\mathrm{argmin}}\, \|\boldsymbol{\xi}\|_1 + \gamma \|\mathbf{y} - \mathbf{W}\boldsymbol{\xi}\|_2^2 \qquad (1.25)$$

for some $\gamma > 0$, or

$$\hat{\mathbf{s}} = \underset{\boldsymbol{\xi} \in \mathbb{R}^n}{\mathrm{argmin}}\, \|\mathbf{y} - \mathbf{W}\boldsymbol{\xi}\|_2^2 \;\text{s.t.}\; \|\boldsymbol{\xi}\|_1 \leq \tau \qquad (1.26)$$

for some $\tau > 0$.

It can also be shown that some γ and τ exist for which (1.25) and (1.26) yield the same solution as BPDN for some noise threshold ε (see, *e.g.*, [98]).

The main limitation of this approach is clearly the choice of γ (or τ), as it strongly influences the outcome of LASSO. A popular criterion involves varying γ in a suitable range to balance the weight between the first and second addend in (1.25) [110].

The problems introduced so far may all be cast into GUROBI by simple manipulations: (1.21) is a linear program, while (1.23) is a quadratically-constrained linear one; (1.26) is a linearly-constrained quadratic program, while (1.25) is an unconstrained quadratic one. These problems are also efficiently solved by *Spectral Projected Gradient for ℓ_1 minimisation (SPGL$_1$)* [111]; such solvers are practically required for tackling large-scale cases with several thousands of variables in a relatively small computation time.

Signal Recovery Algorithms

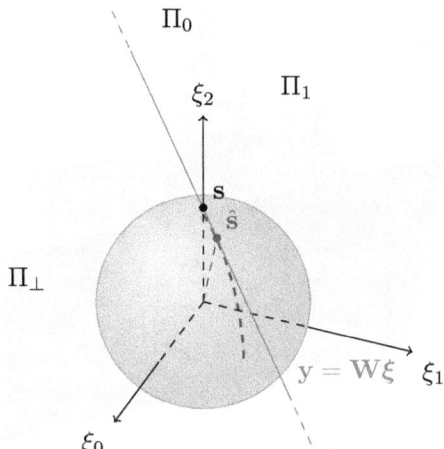

(a) The LLS solution $\hat{\mathbf{s}}$ is found on the smallest ball $\mathcal{B}^\rho_{\ell_2} = \{\boldsymbol{\xi} \in \mathbb{R}^3 : \|\boldsymbol{\xi}\|_2 = \rho\}$ tangent to the flat $\mathbf{y} = \mathbf{W}\boldsymbol{\xi}$, or equivalently at the intersection of the latter with $\Pi_\perp = \mathrm{Ker}(\mathbf{W})^\perp$.

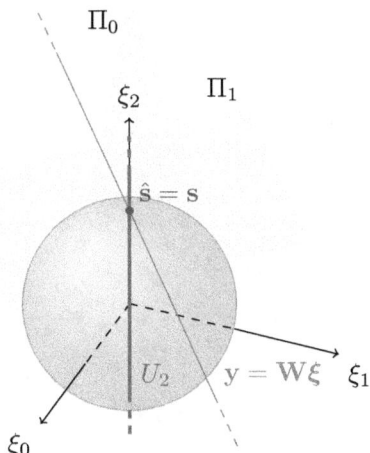

(b) The OLS solution $\hat{\mathbf{s}}$ is found on the smallest ball $\mathcal{B}^\rho_{\ell_2}$ tangent to the intersection between the flat $\mathbf{y} = \mathbf{W}\boldsymbol{\xi}$ and the subspace U_2 (i.e., $T = \{2\}$).

Figure 1.10: Geometric interpretation of the LLS and OLS problems in $\mathbb{R}^3$.

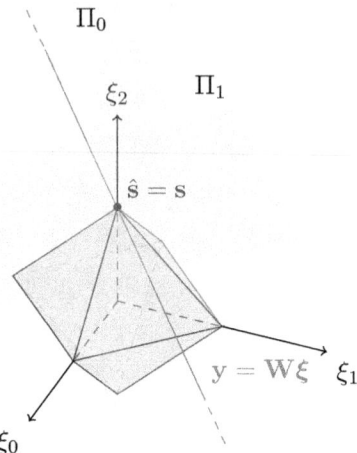

(a) The BP solution $\hat{\mathbf{s}}$ is found on the smallest ball $\mathcal{B}_{\ell_1}^\rho = \{\boldsymbol{\xi} \in \mathbb{R}^3 : \|\boldsymbol{\xi}\|_1 = \rho\}$ tangent to the flat $\mathbf{y} = \mathbf{W}\boldsymbol{\xi}$.

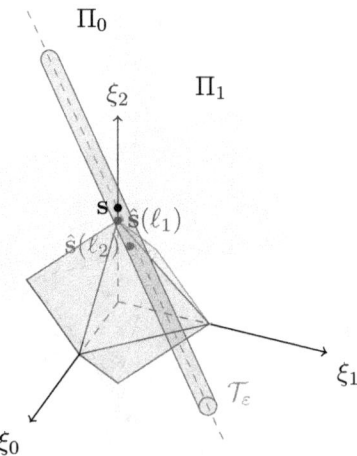

(b) The BPDN solution $\hat{\mathbf{s}}$ is found on the smallest ball $\mathcal{B}_{\ell_1}^\rho$ tangent to the tube $\mathcal{T}_\varepsilon = \{\boldsymbol{\xi} \in \mathbb{R}^3 : \|\mathbf{y} - \mathbf{W}\boldsymbol{\xi}\|_2 \leq \varepsilon\}$. The solution on $\mathcal{B}_{\ell_2}^\rho$ is also reported.

Figure 1.11: Geometric interpretation of the BP and BPDN problems in $\mathbb{R}^3$.

Joint-Sparse and Low-Rank Signal Recovery

The aforementioned problems specifically concentrate on the recovery of a single sparse vector; the straightforward extension to the case of recovering multiple sparse vectors from multiple observations collected as $\mathbf{Y} = \mathbf{WS}, \mathbf{S} \in \mathbb{R}^{n \times w}$ allows one to envision more advanced models that promote the solutions' linear dependence in a convex problem. In fact, any non-convex mixed norm in Definition 1.6 can be replaced by the closest convex mixed (p,q)-norm, e.g., $\|\Xi\|_{2,0}$ can be relaxed to $\|\Xi\|_{2,1}$ to enforce joint-sparsity on a candidate solution $\Xi \in \mathbb{R}^{n \times w}$. With little effort, the similar notion of group sparsity can also be promoted [44].

Similarly to Problem 1.3, finding a low-rank approximation $\hat{\mathbf{X}}$, i.e., minimising the rank of a matrix also leads to an NP-hard problem in general [112, Section 7.3], the cost function $\text{rank}(\Xi)$ being non-convex. However, it is well known that this cost function can be relaxed to the *nuclear norm* $\|\Xi\|_* = \sum_{j=0}^{n-1} \sigma_j(\Xi) = \text{tr}\left(\sqrt{\Xi^*\Xi}\right)$, which is convex. Along with the other relaxations, this adds to an extended set of methods to enforce low-dimensional signal models by convex functions. This general look at signal recovery is given by the following problem.

Problem 1.8 *(Signal Recovery by Convex Optimisation).* The solution $\hat{\mathbf{S}}$ to a generic signal recovery problem with convex objective on a convex set is

$$\hat{\mathbf{S}} = \underset{\Xi \in \mathbb{R}^{n \times w}}{\arg\min} \sum_{i=0}^{P-1} \gamma_i \mathcal{P}_i(\Xi) \text{ s.t. } \bigcap_{i=0}^{C-1} \mathcal{C}_i(\mathbf{Y}, \mathbf{W}, \Xi) \quad (1.27)$$

where $\{\mathcal{P}_i\}_{i=0}^{P-1}$ is a family of convex cost functions of weights $\gamma_i \geq 0$, while $\{\mathcal{C}_i\}_{i=0}^{C-1}$ is a family of convex constraints.

This general problem can be tackled by a number of algorithms; aside from the aforementioned SPGL$_1$ that effectively applies to a subset of such problems, proximal algorithms [96, 97] are a class of efficient iterative solvers with guaranteed convergence for any problem of the form (1.27); a typical example of such methods is Douglas-Rachford splitting [113, 114]. These can be used as solvers once a problem is suitably cast with its standard "resolvent operators" (namely, the *proximity operator* of any objective and *projection operator* of any constraint), their review being left to the given references as lying outside the focus of this thesis.

1.3.2 Recovery Guarantees for Compressed Sensing

We now summarise some results on the recoverability of a single sparse vector from a small number of measurements. These provide a connection between signal recovery algorithms, sensing matrix properties and sparsity. Firstly, a link between the coherence of $\mathbf{W}$ and the recoverability of $\mathbf{s}$ is made.

Theorem 1.7 *(Exact Solution by BP (via coherence) [106])*. Let $\mathbf{y} = \mathbf{Ws}$, $\mathbf{s} \in \Sigma_k$ and $\mathbf{W}$ have coherence $\bar{\mu} < \frac{1}{2k-1}$ (see Theorem 1.6); then $\hat{\mathbf{s}} = \Delta_{\text{BP}}(\mathbf{y}, \mathbf{W})$ is such that $\hat{\mathbf{s}} = \mathbf{s}$.

A celebrated result for BP may then be found [13, 64] by relying on the small k-RIC of the sensing matrix.

Theorem 1.8 *(Exact Solution by BP (via the RIP) [64, Theorem 1.1])*. Let $\mathbf{y} = \mathbf{Ws}$; and $\mathbf{W}$ verify the RIP with constant $\delta_{2k} < \sqrt{2} - 1$. Then $\hat{\mathbf{s}} = \Delta_{\text{BP}}(\mathbf{y}, \mathbf{W})$ is so that

$$\|\hat{\mathbf{s}} - \mathbf{s}\|_2 \leq c_0 \frac{\|\check{\mathbf{s}} - \mathbf{s}\|_1}{\sqrt{k}} \qquad (1.28)$$

for a positive constant

$$c_0 = 2 \frac{1 + (\sqrt{2} - 1)\delta_{2k}}{1 - (\sqrt{2} + 1)\delta_{2k}} \qquad (1.29)$$

Thus, if a signal $\mathbf{x}$ has an exactly k-sparse representation, and once $\mathbf{W}$ has sufficiently large m to ensure that the k-RIC is smaller than $\sqrt{2} - 1$ (e.g., as in (1.14)) exact, lossless signal recovery is possible even if $m < n$, i.e., in an undersampling (or compression, if thought as a coding scheme) regime. When a signal is (k, ϑ)-compressible, (1.28) ensures that the estimate's recovery error norm w.r.t. the solution s is at most as large as set by the residual coefficients w.r.t. its k-sparse approximation. The next Theorem extends (1.28) to evaluate the worst-case recovery error of $\hat{s}$ w.r.t. s when the measurements are affected by additive bounded-energy noise and BPDN is used to decode them.

Theorem 1.9 *(Stable Recovery by BPDN from Noisy Measurements [64, Theorem 1.2])*. Let $\mathbf{y} = \mathbf{W}\mathbf{s} + \boldsymbol{\nu}$ be noisy measurements with additive, independent noise $\boldsymbol{\nu} \in \mathbb{R}^m$ so that $\|\boldsymbol{\nu}\|_2 \leq \varepsilon$; $\mathbf{W}$ verify the RIP with constant $\delta_{2k} < \sqrt{2} - 1$. Then $\hat{\mathbf{s}} = \Delta_{\mathrm{BPDN}}(\mathbf{y}, \mathbf{W}, \varepsilon)$ is so that

$$\|\hat{\mathbf{s}} - \mathbf{s}\|_2 \leq c_0 \frac{\|\check{\mathbf{s}} - \mathbf{s}\|_1}{\sqrt{k}} + c_1 \varepsilon \tag{1.30}$$

for

$$c_0 = 2 \frac{1 + (\sqrt{2} - 1)\delta_{2k}}{1 - (\sqrt{2} + 1)\delta_{2k}}, \quad c_1 = 4 \frac{\sqrt{1 + \delta_{2k}}}{1 + (\sqrt{2} + 1)\delta_{2k}} \tag{1.31}$$

An important extension of this Theorem also accounts for cases in which a perturbation matrix is superimposed to $\mathbf{W}$, and is reviewed in Chapter 5.

Moreover, Theorems 1.8 and 1.9 are based on the RIP, with their error norm bounds depending on the actual value of δ_{2k} in the form of some constants $c_0(\delta_{2k}), c_1(\delta_{2k})$. Nevertheless the conditions provided therein are only sufficient, and (1.30) loosely predicts the typical recovery *Mean-Square Errors (MSEs)* or SNRs as can be numerically verified by simple experiments.

Thus, once a signal is k-sparse w.r.t. some ONB $\mathbf{D}$ and $\mathbf{A}$ is chosen from a RsGE, a minimum value of m can be anticipated from (1.14) so that δ_{2k} of $\mathbf{W}$ is small enough for Theorems 1.8,1.9 to hold. However, such a matrix design is limited by the fact that the RIP considers the worst possible case among all supports of s, regardless of their actual significance in the instance at hand. We therefore stress the need to validate the performances of a CS-based approach to signal acquisition (as in Fig. 1.2) prior to its application by large-scale numerical experiments that ensure accurate signal recovery in typical cases, at least for what concerns the effect of choosing a sensing matrix against another one.

Finally, the performance index that we will use abundantly in the rest of this thesis is $\mathrm{SNR}_{\hat{x},x}$ or $\mathrm{SNR}_{\hat{s},s}$ for single instances; to make some considerations on typical recoveries, we will use the average recovery SNR

$$\mathrm{ASNR}_{\hat{x},x} = 20 \log_{10} \hat{\mathbb{E}} \left[\frac{\|x\|_2}{\|\hat{x} - x\|_2} \right] \mathrm{dB} \quad (1.32)$$

or equivalently

$$\mathrm{ASNR}_{\hat{s},s} = 20 \log_{10} \hat{\mathbb{E}} \left[\frac{\|s\|_2}{\|\hat{s} - s\|_2} \right] \mathrm{dB} \quad (1.33)$$

where the empirical average $\hat{\mathbb{E}}[\cdot]$ is taken over large sets of instances, against which the performances of a particular recovery algorithm, sensing matrix or signal class is tested.

Summary

- ▶ CS is a mathematical framework that allows for a generalised view on the problem of sensing a discrete or continuous signal with an undersampled set of measurements. It leverages the synergy of low-dimensional signal models, sensing operator properties and signal recovery algorithms for which recovery guarantees are provided.

- ▶ Low-dimensional signal models provide low-complexity descriptions of natural and man-made signals. The core principle of sparsity, *i.e.*, the fact that a signal can be decomposed as a linear combination of a few basis elements, can be declined into a number of definitions that quantify how well a signal is represented by a sparse vector in a suitable basis. An extension to multiple instances is also possible; moreover, low-rank models conceptually replace sparsity by the rank of a matrix being recovered. Both models apply to any number of dimensions as shown in a MS imaging example.

- The sensing operators used in CS perform a dimensionality reduction, and cause loss of information in general. A number of properties, such as the RIP and coherence have been introduced with the purpose of verifying if a given configuration of a sensing operator can avoid such information losses w.r.t. a specific signal class, *i.e.*, that of sparse signals. The RMEs that are commonly used in CS were briefly reviewed, as well as some insight on why randomness actually provides a suitable method for dimensionality reduction.

- Some elementary forms of signal recovery algorithms were reviewed, in the specific case of convex optimisation problems. Although not unique, this class of problems and their related solvers allows for an intuitive geometric analogue as to why sparsity can be replaced by a convex proxy. In particular, BP and BPDN will be mentioned as reference decoders several times in this thesis.

- The fundamental guarantees for an exact solution of BP and a bounded-error approximation of the original signal by BPDN were briefly recalled without their technical proof. Their significance is mostly theoretical, as they provide some loose bounds that underestimate the typical performances observed after signal recovery.

I

Adaptive and Efficient Matrix Designs for Compressed Sensing

Maximum Energy Sensing Matrix Designs for Localised Signals

According to the connection between Theorems 1.5, 1.8 a universal design of the sensing matrix $\mathbf{A} \in \mathbb{R}^{m \times n}$ is guaranteed to preserve the information in $\mathbf{x}$ once (i) the sparsity basis or dictionary $\mathbf{D}$ on which $\mathbf{x}$ is supposedly k-sparse does not alter its k-RIC, (ii) $\mathbf{A}$ is drawn from a RsGE whose distribution behaves as close as possible to a RGE, (iii) $m \geq \overline{m}(k,n)$ with $\overline{m}$ essentially determined by the distribution and dimensions of $\mathbf{W}$ as in (1.12).

While this approach holds with no further hypotheses, it does not leverage very simple structures that are often present in most natural signals. In fact, if one looks at a k-sparse signal model there is no explicit mention at how the support or the values of its sparse representation are statistically distributed, or, *e.g.*, whether its most significant coefficients are more likely to appear in some supports rather than others. This has inspired a number of contributions that focus on exploiting similar statistical properties in signal recovery [21, 22, 115]; probabilistic signal models that harness further structure in a sparse or compressible signal are still a subject of intense research [23, 43]. In addition, since the aforementioned fundamental results of CS are only stated for compressible signals, providing actual recovery guarantees based on additional priors is a non-trivial matter.

In this Chapter we aim at adapting a RsGE to the correlation matrix of a signal $\mathbf{C_x}$ (or equivalently of its sparse coefficients $\mathbf{C_s}$); the rationale of this optimisation is *energy maximisation* of the random measurements $\mathbf{y} = \mathbf{Ax}$ (*i.e.*, $\mathbf{y} = \mathbf{Ws}$) as a proxy for their optimality w.r.t. the chosen signal model; this approach stems from [2, 27, 29, 116]. In particular, the evidence presented here shows that (i) natural signals exhibit *localisation*, *i.e.*, the fact that their average energy is spread along a few directions in the signal domain w.r.t. its dimensionality, (ii) maximising the average energy of the random measurements empirically leads to an observable improvement in signal recovery by BP, (iii) sensing matrix designs based on maximising the measurements' average energy may be devised by introducing the concept of *rakeness* and using it to formulate some optimisation problems; these output the optimal correlation matrix w.r.t. this criterion for the rows of $\mathbf{W}$, that is therefore generated as a RME with non-i.i.d. entries. Finally, some considerations on the RIP of such matrices illustrate the specificity of such an adaptation. This treatment is completed by numerical evidence on the phase transition of BP w.r.t. maximum energy sensing matrices as applied on sparse and localised signals.

2.1 Sparse and Localised Signals

Deterministic k-sparse signal models essentially make no assumptions on the statistical properties of either $\mathbf{x}$ or $\mathbf{s}$ once they are considered as R.V.s from a suitable probability space[1]. Without further priors, this could conform to a case in which each of $\binom{n}{k}$ supports for $\|\mathbf{s}\|_0 = k$ has equal probability; nevertheless, this rarely occurs when the signal of interest carries information content. We now discuss a simple simulation model for vectors that are sparse and non-white.

---

[1]With slight abuse, random vectors [117] and specific instances share the same notation and disambiguation is made explicit where needed.

2.1.1 A Generative Model for Synthetic Sparse and Non-White Signals

In general, the problem of defining a PDF that generates k-sparse vectors is non-trivial [23, 118]; the purpose of this Section is to define a PDF associated with a k-sparse and non-white R.V. s, *i.e.*, so that such a vector can be generated for simulation purposes.

To do so, we let the *Probability Mass Function (PMF)* of a support $T \subseteq \{0, \ldots, n-1\}$ be

$$\mathbb{P}[T = \text{supp}(\mathbf{s})] = \gamma_T, \quad \sum_{\substack{T \subseteq \{0,\ldots,n-1\} \\ |T|=k}} \gamma_T = 1$$

For a given value of T, $\mathbf{s}_T$ is then taken as a R.V. of $\mathbb{R}^k$ with suitable PDF; this method corresponds to generating a *multivariate mixture*

$$\mathbf{s} \sim \sum_{\substack{T \subseteq \{0,\ldots,n-1\} \\ |T|=k}} \gamma_T \, \delta(\mathbf{s}_{T^c}) \, f(\mathbf{s}_T) \qquad (2.1)$$

where $\delta(\mathbf{s}_{T^c})$ is the Dirac delta in $\mathbb{R}^{n-k}$ and $f(\mathbf{s}_T)$ is the PDF of $\mathbf{s}_T$ (*i.e.*, conditioned to the support T). Moreover, if we let all $\mathbf{s}_T$ have mean $\boldsymbol{\mu}_{\mathbf{s}_T} = \mathbf{0}_k$, the covariance of s is

$$\mathbf{K}_\mathbf{s} = \sum_{T \subseteq \{0,\ldots,n-1\}, |T|=k} \gamma_T \mathbf{K}_{\mathbf{s}_T}$$

As an example, a common way of generating a k-sparse vector takes $\gamma_T = \binom{n}{k}^{-1}$, $\mathbf{s}_T \sim \mathcal{N}(\mathbf{0}_k, \mathbf{K}_{\mathbf{s}_T})$ where the covariance matrix $\mathbf{K}_{\mathbf{s}_T} = \frac{1}{k}\mathbf{I}_k$; thus (2.1) generates k-sparse vectors whose covariance $\mathbf{K}_\mathbf{s} = \binom{n}{k}^{-1} \frac{\binom{n-1}{k-1}}{k} \mathbf{I}_n = \frac{1}{n}\mathbf{I}_n$, *i.e.*, this choice generates a white R.V. s.

With the exception of the latter case, (2.1) is impractical to implement for a large number of supports, yet it serves the purpose of showing how both the support and coefficient distributions of s can be controlled, at least with the aim of altering the second-order moments in $\mathbf{K}_\mathbf{s}$: any significant change in the probability assignment γ_T or covariance $\mathbf{K}_{\mathbf{s}_T}$ for some T will lead to a non-white k-sparse R.V., opening the possibility of simulating it.

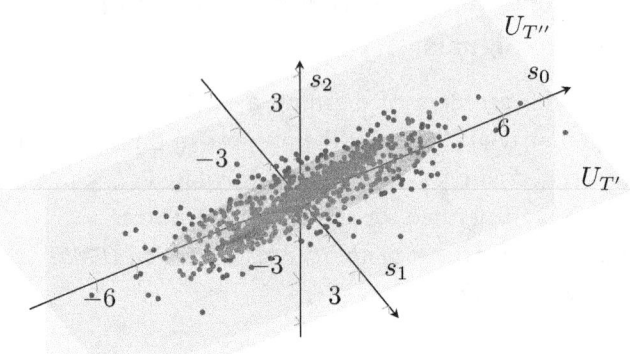

Figure 2.1: A centred multivariate Gaussian mixture defined on two subspaces $U_{T'}, U_{T''}$: samples in $T' = \{0,1\}$ (blue dots) and $T'' = \{0,2\}$ (purple dots); the ellipsoid $\{s \in \mathbb{R}^n : s^*\mathbf{K_s}^{-1}s = c\}$ for some $c > 0$ (overlay).

This approach merely attempts to emulate what happens in natural signals that verify a sparse signal model in each of their instances, yet maintain a non-white correlation matrix. Thus, the simulation method discussed here may be avoided in favour of using real-world datasets, since the adaptation discussed in this Chapter requires the knowledge of $\mathbf{K_s}$ (or $\mathbf{C_s}$).

2.1.2 Localisation

We now move the focus to discussing the correlation matrix of a R.V. s regardless of its sparsity. This symmetric *Positive-Semidefinite (PSD)* matrix contains valuable information on it: by the spectral theorem for symmetric matrices $\mathbf{C_s} \stackrel{\text{sed}}{=} \mathbf{U_s \Lambda_s U_s^*}$ the n orthonormal eigenvectors in $\mathbf{U_s}$ describe how the R.V. is distributed in $\mathbb{R}^n$. In particular, if s follows an elliptically-contoured distribution[2] $\mathbf{C_s}$ is directly related to the shape of the level curves of $f(s)$; if not, the eigenvectors in $\mathbf{U_s}$ simply indicate the directions along which s spreads out on-average. For this

---

[2]In a simplified fashion, whose PDF $f(s) \propto g(s^*\mathbf{K_s}^{-1}s)$ for g verifying some integrability conditions [119].

geometric intuition it is common to adopt the following terminology: we define zero-mean (*i.e., centred*) white R.V.s as *isotropic*, and any non-centred or non-white R.V. as *anisotropic*.

To fix ideas, one could consider a centred k-sparse vector s distributed as (2.1) over two possible supports with probability $\gamma_{T'} = \gamma_{T''} = \frac{1}{2}$, setting the others to 0. An instance of such a vector would then occupy two subspaces $U_{T'}, U_{T''}$ and have mixture covariance $\mathbf{K_s} = \frac{1}{2}\left(\mathbf{K}_{\mathbf{s}_{T'}} + \mathbf{K}_{\mathbf{s}_{T''}}\right)$. In the case of a centred Gaussian mixture this example is reported in Fig. 2.1 where it is visually appreciated that the distribution is strongly anisotropic, as the depicted instances of s are indeed on-average concentrated along the eigenvector corresponding to $\lambda_{\max}(\mathbf{K_s})$. In this context we introduce the following definition to quantify the anisotropy of s.

Definition 2.1 (*Localisation [2]*). Let $\mathbf{s} \in \mathbb{R}^n$ be a R.V. with correlation matrix $\mathbf{C_s} \stackrel{\text{sed}}{=} \mathbf{U_s \Lambda_s U_s^*}$ of eigenvalues $\mathbf{\Lambda_s}$ and average energy $\mathcal{E}_\mathbf{s} = \operatorname{tr}(\mathbf{C_s})$. We define *localisation* as

$$\mathcal{L}(\mathbf{s}) = \operatorname{tr}\left[\left(\frac{\mathbf{\Lambda_s}}{\mathcal{E}_\mathbf{s}} - \frac{1}{n}\mathbf{I}_n\right)^2\right] \tag{2.2}$$

Thus $\mathcal{L}(\mathbf{s})$ simply measures a deviation in the suitably normalised eigenvalues of $\mathbf{C_s}$ from the white case. Note that $\mathcal{L}(\mathbf{s}) \in [0, \frac{n-1}{n}]$, *i.e.*, minimum localisation corresponds to the white case, *i.e.*, $\forall j \in \{0, \ldots, n-1\}$, $\lambda_j(\mathbf{C_s}) = \frac{\mathcal{E}_\mathbf{s}}{n}$, while maximum localisation is obtained when $\operatorname{rank}(\mathbf{C_s}) = 1$, that is when $\lambda_{\max}(\mathbf{C_s}) = \lambda_0(\mathbf{C_s}) = \mathcal{E}_\mathbf{s}$ and $\forall j \in \{1, \ldots, n-1\}$, $\lambda_j(\mathbf{C_s}) = 0$.

To see if natural signals are localised, we let $\mathbf{D}$ be an ONB and $\mathbf{x} = \mathbf{Ds}$, so $\mathcal{L}(\mathbf{x}) = \mathcal{L}(\mathbf{s})$. Thus, we may evaluate their localisation by populating a matrix $\mathbf{X} \in \mathbb{R}^{n \times w}$ with a large number w of instances of the R.V. $\mathbf{x}$, and estimate quite simply the sample correlation matrix as $\widehat{\mathbf{C}}_\mathbf{x} = \frac{1}{w}\mathbf{X}\mathbf{X}^*$, which is then used to compute $\mathcal{L}(\mathbf{x})$. Applying this procedure to a variety of *Electrocardiographic Tracks (ECGs)* and *Electromyographic Tracks (EMGs)* from [120], as well as 10 ms-long speech segments from [121] and greyscale images of letters generated as in [29] yields the values reported in Table 2.1.

Signal Class	Sampling Rate	n	$\mathcal{L}(\mathbf{x})$
ECG	720 Hz	360	0.187
Speech segments	20 KHz	200	0.069
EMG	400 Hz	200	0.021
Grayscale printed letters	24×24 pixel	576	0.016

Table 2.1: Estimation of $\mathcal{L}(\mathbf{x})$ for some real-world signal classes.

From this empirical evaluation we may safely conclude that natural signals are indeed localised according to Definition 2.1, and that when $\mathbf{C_x}$ (or $\mathbf{C_s}$) is a stationary property of n-dimensional R.V.s extracted from the process that generates $\mathbf{x}$ an adaptation of the sensing matrix w.r.t. this feature is indeed possible, as explained in the next Section.

2.2 Rakeness and the Rationale of Energy Maximisation

The problem of extracting the maximum amount of information from a k-sparse and localised R.V. s involves defining (i) a figure of merit to evaluate the RME from which $\mathbf{W}$ is drawn as a sensing operator and (ii) a verification that once the chosen figure of merit attains its optimal value, this corresponds to an improvement in terms of MSE (*i.e.*, of $\mathrm{SNR}_{\hat{s},s}$) attained by a suitable signal recovery algorithm. We here discuss the choice of *rakeness* as this figure.

2.2.1 Rakeness

The definition of rakeness was developed in a number of contributions [2, 27–29] to reach the following forms.

Definition 2.2 (*Rakeness (single R.V. case [29])*). Let $s, w \in \mathbb{R}^n$ be independent R.V.s with correlation matrices $\mathbf{C_s}$ and $\mathbf{C_w}$ respectively.

We define *rakeness* as[a]

$$\mathcal{R}_\mathbf{s}(\mathbf{w}) = \mathbb{E}_{\mathbf{w},\mathbf{s}}[|\mathbf{w}^*\mathbf{s}|^2] = \sum_{j=0}^{n-1}\sum_{l=0}^{n-1}(\mathbf{C}_\mathbf{w} \circ \mathbf{C}_\mathbf{s})_{j,l} = \operatorname{tr}(\mathbf{C}_\mathbf{w}\mathbf{C}_\mathbf{s}) \quad (2.3)$$

Note that (2.3) amounts to the *Frobenius scalar product* between $\mathbf{C}_\mathbf{w}$ and $\mathbf{C}_\mathbf{s}$. The above definition is extended to a set of m R.V.s as follows.

Definition 2.3 (*Rakeness (multiple R.V. case)*). Let $\mathbf{s} \in \mathbb{R}^n$ be a R.V. with correlation matrix $\mathbf{C}_\mathbf{s}$; $\mathbf{W} \in \mathbb{R}^{m \times n}$ be a RME where each row is drawn from a different R.V. in $\{\mathbf{w}_i\}_{i=0}^{m-1}$ with correlation matrix $\mathbf{C}_{\mathbf{w}_i}$; $\mathbf{W}$ and $\mathbf{s}$ be independent. We define *rakeness* as

$$\mathcal{R}_\mathbf{s}(\mathbf{W}) = \mathbb{E}_{\mathbf{W},\mathbf{s}}[\|\mathbf{W}\mathbf{s}\|_2^2] = \sum_{i=0}^{m-1}\mathcal{R}_\mathbf{s}(\mathbf{w}_i) \quad (2.4)$$

Thus, (2.3) measures the average energy extracted from $\mathbf{s}$ by means of linear projection over another R.V.. Clearly, Definition 2.2 applies to evaluating this quantity for a single R.V. or a sensing matrix $\mathbf{W}$ drawn from a RME with i.i.d. rows. Definition 2.3 merely extends the same concept to RMEs with non-i.i.d. rows.

2.2.2 The Rationale of Energy Maximisation

We have previously mentioned the role of the RIP in CS as a method to ensure that the pairwise distances between any two $\mathbf{s}', \mathbf{s}'' \in \Sigma_k$ are preserved. Clearly, the essential purpose of this property is that $\|\mathbf{W}(\mathbf{s}' - \mathbf{s}'')\|_2^2 = 0$ never occurs for any two k-sparse vectors – a complete aliasing of their images $\mathbf{y}' = \mathbf{y}''$ would otherwise imply no recovery algorithm is able to retrieve $\mathbf{s}', \mathbf{s}''$. In fact, an asymmetric formulation of the RIP takes this into account [73] and distinguishes between the lower and upper bound of (1.8); as a partial guarantee that $\mathbf{y}' \neq \mathbf{y}''$ for $\mathbf{s}', \mathbf{s}'' \in \Sigma_k$, we could require that

$$\mathbb{P}\left[\|\mathbf{W}(\mathbf{s}' - \mathbf{s}'')\|_2^2 \geq \zeta\|\mathbf{s}' - \mathbf{s}''\|_2^2\right] \simeq 1, \; \zeta \in (0,1] \quad (2.5)$$

---

[a] The final form uses the identity $\mathbf{1}_n^*(\mathbf{U} \circ \mathbf{V})\mathbf{1}_n = \operatorname{tr}(\mathbf{U}\mathbf{V}^*)$ [59, Section 5.7].

for some ζ as close to 1 as possible. Rather than a probability bound for sparse vectors in Σ_k, we further approximate the concept of (2.5) as follows. Assuming that $\mathbf{s}'$ and $\mathbf{s}''$ are independent copies of a R.V. $\mathbf{s}$ with mean $\boldsymbol{\mu}_\mathbf{s}$ and covariance $\mathbf{K}_\mathbf{s}$ we evaluate $\mathbb{E}_{\mathbf{W},\mathbf{s}',\mathbf{s}''}\left[\|\mathbf{W}(\mathbf{s}'-\mathbf{s}'')\|_2^2\right]$. By expanding this in the case of a centred RME $\mathbf{W}$ whose rows have correlations and covariances both equal to $\{\mathbf{C}_{\mathbf{w}_i}\}_{i=0}^{n-1}$, we see that

$$\mathbb{E}_{\mathbf{W},\mathbf{s}',\mathbf{s}''}\left[\|\mathbf{W}(\mathbf{s}'-\mathbf{s}'')\|_2^2\right] = 2\left(\mathcal{R}_\mathbf{W}(\mathbf{s}) - \sum_{i=0}^{m-1}\boldsymbol{\mu}_\mathbf{s}^*\mathbf{C}_{\mathbf{w}_i}\boldsymbol{\mu}_\mathbf{s}\right) \quad (2.6)$$

Clearly (2.5) is a much stronger statement to verify, yet (2.6) communicates that

$$\mathbb{E}_{\mathbf{W},\mathbf{s}',\mathbf{s}''}\left[\|\mathbf{W}(\mathbf{s}'-\mathbf{s}'')\|_2^2\right] \propto \sum_{i=0}^{m-1}\mathcal{R}_\mathbf{s}(\mathbf{w}_i)$$

so the more rakeness is made large, the higher the average energy attained by the measurements. Thus, maximising rakeness (2.6) by acting on $\mathbf{W}$ could indeed mitigate (in expectation) the chance that the images of two $\mathbf{s}', \mathbf{s}''$ that comply with the hypothesis have Euclidean distance significantly smaller than $\|\mathbf{s}'-\mathbf{s}''\|_2$. However, since no explicit relation to sparsity is made here, one could doubt that a maximisation of the measurements' energy (as implied by maximising rakeness as in (2.6)) has little effect on the recoverability of $\mathbf{s}$; or, perhaps, that it could even limit the performances of a recovery algorithm.

We now propose reassuring evidence that maximising the measurements' energy does improve signal recovery; to show this, we perform a numerical experiment that does not rely on localisation, but on a simple *a posteriori* selection of sets of measurements with maximum energy. We consider R.V.s $\mathbf{s}$ with k-sparse realisations, having i.i.d. supports T and generated as $\mathbf{s}_T \sim \mathcal{N}(\mathbf{0}_k, \frac{1}{k})$. We let $\underline{\mathbf{W}} \in \mathbb{R}^{m \times n}$ be drawn from the RGE and collect $\underline{\mathbf{y}} = \underline{\mathbf{W}}\mathbf{s}$, where $n = 256$, $\underline{m} = 10^4$, i.e., we collect a large number of Gaussian random measurements. On these, we apply two *a posteriori* selections: we either (i) take $\mathbf{y}' \in \mathbb{R}^m$ as the first m components of $\underline{\mathbf{y}} \in \mathbb{R}^{\underline{m}}$ (thus reproducing the standard setting); or (ii) we use the m largest components of $\underline{\mathbf{y}}$, thus maximising the energy of the corresponding $\mathbf{y}'' \in \mathbb{R}^m$.

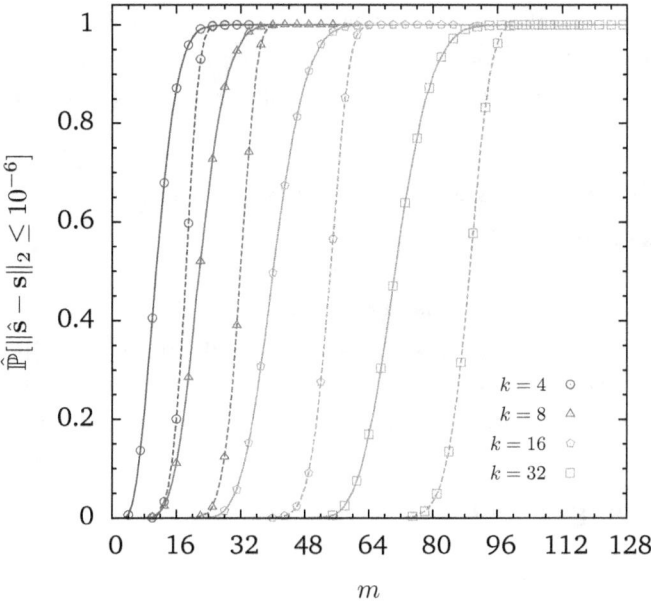

Figure 2.2: Probability of successful recovery for maximum energy $\mathbf{y}''$ (solid lines) *versus* classic $\mathbf{y}'$ (dotted lines) at different sparsity levels.

Two reconstructions of s are then performed by solving BP with CPLEX [94], depending on whether (1.21) is solved w.r.t. $\mathbf{y}'$ or $\mathbf{y}''$. The probability of successful recovery of s will indeed depend on $k = \|\mathbf{s}\|_0$ and m. The results are reported in Fig. 2.2 in terms of the empirical probability $\hat{\mathbb{P}}[\|\hat{\mathbf{s}} - \mathbf{s}\|_2 \leq 10^{-6}]$ as a function of $k = \{4, 8, 16, 32\}$, the number of measurements $m \in \{1, \ldots, \frac{n}{2}\}$, and the choice of $\mathbf{y}', \mathbf{y}''$. We emphasise that the latter vectors are both noiseless, same-dimensionality observations of the same s with $\mathcal{L}(\mathbf{s}) = 0$.

The evidence indicates that, even in a noiseless configuration and with a white R.V. s, choosing *a posteriori* the measurements with the largest energy moves $\hat{\mathbb{P}}[\|\hat{\mathbf{s}} - \mathbf{s}\|_2 \leq 10^{-6}] = \frac{1}{2}$ to lower m. Although this is not a rigorous proof and this toy example has no practical advantage (as it relies on selecting the best m out of $10^4 \gg n$ projections), we still have a clear indication that increasing the measurements' energy enhances the probability of successful signal recovery, provided that k

is sufficiently small. Since rakeness is a proxy for the measurements' average energy w.r.t. localised signals, its maximisation is expected to have a beneficial effect on the recoverability of k-sparse signals.

2.3 Maximum Energy Sensing Matrix Designs for Compressed Sensing

The next step is ensuring that a behaviour such as that of Fig. 2.2 can be obtained by maximising (2.4), *i.e.*, by adapting the second-moments of $\mathbf{W}$ to those of s. Thus, we approach the formulation of an *energy maximisation* problem in which $\mathcal{L}(\mathbf{s})$ plays a fundamental role.

2.3.1 Energy Maximisation Problems

The i.i.d. rows case

We first introduce an optimisation problem that yields a covariance matrix $\mathbf{C_w}$ maximising (2.3), therefore producing a requirement on a R.V. w by which, *e.g.*, RMEs with i.i.d. rows can be designed.

Problem 2.1 (*Maximum Energy Sensing Matrix Design (i.i.d. rows case)*). Let $\mathbf{s}, \mathbf{w} \in \mathbb{R}^n$ be independent R.V.s with correlation matrices $\mathbf{C_s} \stackrel{\text{sed}}{=} \mathbf{U_s \Lambda_s U_s^*}$ and $\mathbf{C_w} \stackrel{\text{sed}}{=} \mathbf{U_w \Lambda_w U_w^*}$ respectively. A design of w that maximises (2.3) is obtained by letting

$$\mathbf{C_w}^{(\tau)} = \underset{\mathbf{C_w} \succeq 0}{\operatorname{argmax}} \, \mathcal{R}_\mathbf{w}(\mathbf{s}) \tag{2.7}$$

s.t.

$$\mathcal{E}_\mathbf{w} = 1 \tag{2.8}$$

$$\mathcal{L}(\mathbf{w}) \leq \tau \mathcal{L}(\mathbf{s}) \tag{2.9}$$

with the whiteness parameter $\tau \in [0, 1]$.

This seemingly complicated optimisation problem has an extremely simple closed-from solution, as detailed in the following Proposition.

Proposition 2.1 *(Eigenvalue form of Problem 2.1)*. The solution of Problem 2.1 is equivalent to $\mathbf{C}_\mathbf{w}^{(\tau)} \stackrel{\text{sed}}{=} \mathbf{U}_\mathbf{s} \mathbf{\Lambda}_\mathbf{w}^{(\tau)} \mathbf{U}_\mathbf{s}^*$ where

$$\mathbf{\Lambda}_\mathbf{w}^{(\tau)} = \underset{\mathbf{\Lambda}_\mathbf{w} \succeq 0}{\operatorname{argmax}} \operatorname{tr}(\mathbf{\Lambda}_\mathbf{w} \mathbf{\Lambda}_\mathbf{s}) \qquad (2.10)$$

s.t.

$$\mathcal{E}_\mathbf{w} = 1 \qquad (2.11)$$

$$\operatorname{tr}(\mathbf{\Lambda}_\mathbf{w}^2) \leq \tau \frac{\operatorname{tr}(\mathbf{\Lambda}_\mathbf{s}^2)}{\mathcal{E}_\mathbf{s}^2} + \frac{1 - \tau}{n} \qquad (2.12)$$

We now illustrate how each part of Problem 2.1 and its eigenvalue form may be derived.

Proof of Problem 2.1. We begin by recalling that $\mathbf{C}_\mathbf{w}, \mathbf{C}_\mathbf{s}$ are by definition symmetric PSD, so $\mathbf{C}_\mathbf{w} \stackrel{\text{sed}}{=} \mathbf{U}_\mathbf{w} \mathbf{\Lambda}_\mathbf{w} \mathbf{U}_\mathbf{w}^*, \mathbf{C}_\mathbf{s} \stackrel{\text{sed}}{=} \mathbf{U}_\mathbf{s} \mathbf{\Lambda}_\mathbf{s} \mathbf{U}_\mathbf{s}^*$ with $\mathbf{U}_\mathbf{w}, \mathbf{U}_\mathbf{s}$ both unitary matrices containing the eigenvectors of $\mathbf{C}_\mathbf{w}, \mathbf{C}_\mathbf{s}$. In this setting, the Wielandt-Hoffman inequality [122] (reported as [59, Theorem 4.3.53]) grants that

$$\operatorname{tr}(\mathbf{C}_\mathbf{s} \mathbf{C}_\mathbf{w}) \leq \operatorname{tr}(\mathbf{\Lambda}_\mathbf{s} \mathbf{\Lambda}_\mathbf{w}) \qquad (2.13)$$

where equality (2.13) is only attained when $\mathbf{U}_\mathbf{w} = \mathbf{U}_\mathbf{s}$. Thus, starting from the unconstrained, unbounded problem

$$\mathbf{C}_\mathbf{w}^{(\tau)} = \underset{\mathbf{C}_\mathbf{w} \succeq 0}{\operatorname{argmax}} \mathcal{R}_\mathbf{w}(\mathbf{s})$$

$$= \underset{\substack{\mathbf{U}_\mathbf{w} \in \mathbb{R}^n : \mathbf{U}_\mathbf{w}^* \mathbf{U}_\mathbf{w} = \mathbf{I}_n \\ \mathbf{\Lambda}_\mathbf{w} \succeq 0}}{\operatorname{argmax}} \operatorname{tr}(\mathbf{U}_\mathbf{s} \mathbf{\Lambda}_\mathbf{s} \mathbf{U}_\mathbf{s}^* \mathbf{U}_\mathbf{w} \mathbf{\Lambda}_\mathbf{w} \mathbf{U}_\mathbf{w}^*)$$

$$= \underset{\substack{\mathbf{C}_\mathbf{w} \stackrel{\text{sed}}{=} \mathbf{U}_\mathbf{s} \mathbf{\Lambda}_\mathbf{w} \mathbf{U}_\mathbf{s}^* \\ \mathbf{\Lambda}_\mathbf{w} \succeq 0}}{\operatorname{argmax}} \operatorname{tr}(\mathbf{\Lambda}_\mathbf{w} \mathbf{\Lambda}_\mathbf{s}) \qquad (2.14)$$

where we have let $\mathbf{U}_\mathbf{w} = \mathbf{U}_\mathbf{s}$. Since the constraints (2.8),(2.9) do not affect the choice of $\mathbf{U}_\mathbf{w}$ the latter form of (2.14) is equivalent to the objective (2.7), and Problem 2.1 may be fully recast in terms of the eigenvalues in the diagonal matrix $\mathbf{\Lambda}_\mathbf{w} \succeq 0$. The constraint (2.8) (or (2.11)) is then added to fix the average energy $\mathcal{E}_\mathbf{w} = \operatorname{tr}(\mathbf{\Lambda}_\mathbf{w}) = 1$, thus making (2.14) bounded.

Without further constraints, it can be simply shown by the *Karush-Kuhn-Tucker conditions (KKT)* [95, Section 5.5.3] that the solution would be letting $\lambda_{\max}(\mathbf{C_w}) = 1, \lambda_j(\mathbf{C_w}) = 0, j \in \{1,\ldots,n-1\}$, *i.e.*, rakeness is maximised by projecting s along the eigenvector in $\mathbf{U_w}$ corresponding to $\lambda_{\max}(\mathbf{C_w})$. To prevent this over-tuning and preserve the information distributed along all eigenvectors associated to non-zero eigenvalues in $\mathbf{C_s}$ we introduce a localisation constraint on $\mathbf{w}$ as $\mathcal{L}(\mathbf{w}) \leq \tau \mathcal{L}(\mathbf{s})$ yielding (2.12); this behaviour is regulated by a factor $\tau \in [0,1]$ in (2.9), (2.12). $\square$

An intuitive geometric interpretation of this optimisation problem is that $\mathbf{C_w}$ and $\mathbf{C_s}$ define centred ellipsoids of $\mathbb{R}^n$, and maximising the trace in (2.7) with $\mathbf{C_s}$ fixed is indeed a problem of rotating the principal axes defined by $\mathbf{U_w}$ and altering the semi-axis lengths in $\mathbf{\Lambda_w} = \operatorname{diag}\left(\{\lambda_j(\mathbf{C_w})\}_{j=0}^{n-1}\right)$ so that the two centred ellipsoids are aligned. The constraints simply impose that this alignment is normalised, and that the eccentricity of the ellipsoid corresponding to $\mathbf{C_w}$ is always less than that of $\mathbf{C_s}$.

Even if Problem 2.1 has been verbosely presented, we are substantially dealing with a simple quadratically-constrained linear program in n variables; a closed-form solution exists, as shown below.

Proposition 2.2 (*Closed-form Solution of Problem 2.1*). Problem 2.1 is solved by letting $\mathbf{C}_\mathbf{w}^{(\tau)} \stackrel{\text{sed}}{=} \mathbf{U_s \Lambda_w U_s^*}$ where

$$\mathbf{\Lambda_w} = \sqrt{\tau}\frac{\mathbf{\Lambda_s}}{\mathcal{E}_\mathbf{s}} + \frac{1-\sqrt{\tau}}{n}\mathbf{I}_n \qquad (2.15)$$

for $\tau \in [0,1]$. In particular, there exists $\tau^* = \left(1 - \frac{n\lambda_{\min}(\mathbf{C_s})}{\mathcal{E}_\mathbf{s}}\right)^{-2}$ so that for $\tau < \tau^*$, (2.15) is strictly positive.

Without further constraints, Problem 2.1 is simply solved by a linear combination between $\mathbf{\Lambda_s}$ and the normalised identity matrix $\mathbf{I}_n/n$.

Proof of Proposition 2.2. The proof of (2.15) follows by simple, standard application of the KKT to the eigenvalue form of Problem 2.1. τ^*

follows by setting $\forall j \in \{0,\ldots,n-1\}, (\Lambda_{\mathbf{w}})_{j,j} > 0$ and checking the one corresponding to the smallest eigenvalue in $\Lambda_{\mathbf{s}}$. □

The independent, non-identically distributed rows case

An extension of Problem 2.1 to the more general case of a RME with independent, non-i.i.d. rows follows.

Problem 2.2 *(Maximum Energy Sensing Matrix Design (general case))*. Let $\mathbf{s} \in \mathbb{R}^n$ be a R.V. with correlation matrix $\mathbf{C_s}$; let $\mathbf{W} \in \mathbb{R}^{m \times n}$ be a RME with $m \leq n$ independent rows, each drawn from a different R.V. in $\{\mathbf{w}_i\}_{i=0}^{m-1}$ with correlation matrices $\{\mathbf{C}_{\mathbf{w}_i}\}_{i=0}^{m-1}$. A design of $\mathbf{W}$ that maximises (2.4) is obtained by letting

$$\{\mathbf{C}_{\mathbf{w}_i}^{(\tau)}\}_{i=0}^{m-1} = \operatorname*{argmax}_{\{\mathbf{C}_{\mathbf{w}_i}\}_{i=0}^{m-1}, \mathbf{C}_{\mathbf{w}_i} \succeq 0} \mathcal{R}_{\mathbf{W}}(\mathbf{s}) \quad (2.16)$$

s.t.

$$\forall i \in \{0,\ldots,m-1\}, \quad \mathcal{E}_{\mathbf{w}_i} = 1 \quad (2.17)$$
$$\mathcal{L}(\mathbf{w}_i) \leq \tau \mathcal{L}(\mathbf{s}) \quad (2.18)$$
$$\forall (i,j) \in \{0,\ldots,m-1\}^2, \mathcal{R}_{\mathbf{w}_i}(\mathbf{w}_j) \leq \gamma \quad (2.19)$$

with the whiteness parameter $\tau \in [0, +\infty)$ and the similarity parameter $\gamma \in [0, 1]$.

Problem 2.2 is a semidefinite programming problem [112]; its derivation is analogous to Problem 2.1, which is actually a particular case.

Proposition 2.3 *(Particular cases of Problem 2.2)*. Consider any instance of Problem 2.2. Then

- for $\gamma = 1, \tau \in [0,1]$ the constraint (2.19) is inactive, and the problem can be separated in m identical sub-problems that are instances of Problem 2.1, i.e., $\forall i \in \{0,\ldots,m-1\}, \mathbf{C}_{\mathbf{w}_i}^{(\tau)} = \mathbf{C}_{\mathbf{w}}^{(\tau)}$ solved as in (2.15);

- for $\gamma = 0, \tau = +\infty$ the constraint (2.19) imposes $\mathbb{E}\left[|\mathbf{w}_i^* \mathbf{w}_j|^2\right] =$

0, *i.e.*, the m R.V.s by which $\mathbf{W}$ is drawn have zero inner product in expectation. The optimal solution follows by assuming $\forall i \in \{0, \ldots, m-1\}$, $\mathbf{C}_{\mathbf{w}_i} = \mathbf{u}_i \mathbf{u}_i^*$ with $\mathbf{u}_i \in \mathbb{R}^n$ an eigenvector in $\mathbf{U}_{\mathbf{s}}$ corresponding to the i-th largest $\lambda_i(\mathbf{C}_{\mathbf{s}})$.

Although no closed-form solution is discussed here, we conjecture that this problem can be written in a semidefinite programming solver to deliver a set of m R.V.s that, aside from maximising (2.4), verify a constraint on their similarity. In practice, since there is no evidence that introducing (2.19) has a beneficial effect on signal recovery, we will assume $\gamma = 1$, and synthesise R.V.s with the correlation matrix obtained from (2.15).

2.3.2 Synthesis of Maximum Energy Random Sensing Matrices

We now define two RMEs that lend themselves to producing realisations having a given correlation matrix $\mathbf{C}_{\mathbf{w}}$.

Definition 2.4 *(Anisotropic Random Gaussian Ensemble).* We define Anisotropic Random Gaussian Ensemble a RME $\mathbf{W} \in \mathbb{R}^{m \times n}$ whose i.i.d. rows are copies of a R.V. $\mathbf{w} \sim \mathcal{N}(\mu_{\mathbf{w}}, \mathbf{K}_{\mathbf{w}})$ with correlation matrix $\mathbf{C}_{\mathbf{w}} = \mathbf{K}_{\mathbf{w}} + \mu_{\mathbf{w}} \mu_{\mathbf{w}}^*$.

The synthesis method that allows the generation of a $\mathbf{W}$ implementing the $\mathbf{C}_{\mathbf{w}}^{(\tau)}$ resulting from Problem 2.1 is straightforward for *Anisotropic Random Gaussian Ensembles (aRGEs)*.

Proposition 2.4 *(Synthesis of an aRGE).* An aRGE with row mean $\mu_{\mathbf{w}} = \mathbf{0}_n$, row correlation matrix $\mathbf{C}_{\mathbf{w}}$ may be synthesised from a RGE $\mathbf{T} \in \mathbb{R}^{m \times n}$ by letting

$$\mathbf{W} = \mathbf{T} \, (\mathbf{C}_{\mathbf{w}})^{\frac{1}{2}} \qquad (2.20)$$

While the synthesis of an aRGE is extremely simple (see the scheme in Fig 2.3a), the following ensemble is significantly more appealing as it uses only two antipodal symbols and is therefore more favourable in

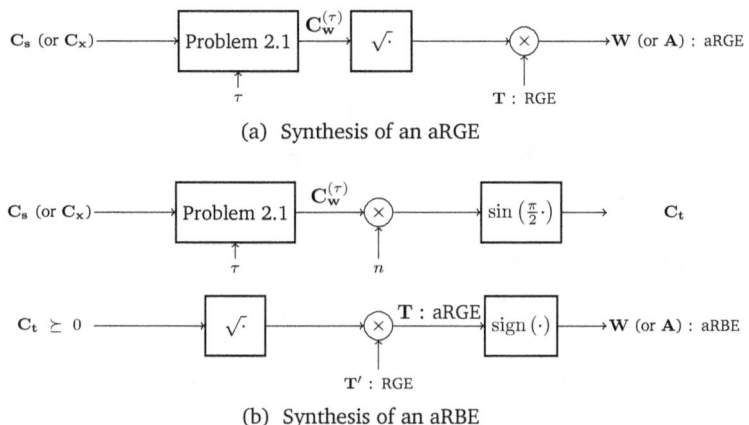

Figure 2.3: Synthesis of maximum energy random sensing matrices.

the perspective of devising a convenient analog or digital implementation.

Definition 2.5 *(Anisotropic Random Bernoulli Ensemble).* We define Anisotropic Random Bernoulli Ensemble a RME $\mathbf{W} \in \mathbb{R}^{m \times n}$ whose i.i.d. rows are copies of a R.V. $\mathbf{w} \in \{-1, +1\}^n$ with correlation matrix $\mathbf{C_w} = \mathbf{K_w} + \mu_\mathbf{w} \mu_\mathbf{w}^* : \forall j \in \{0, \ldots, n-1\}, (\mathbf{C_w})_{j,j} = 1$. A normalisation factor on $\mathbf{W}$ may also be considered.

The problem of synthesising an *Anisotropic Random Bernoulli Ensemble* (aRBE) with a given $\mathbf{C_w}$ is non-trivial and spans a number of existing contributions which cover some special cases [123–125]; clearly, not all $\mathbf{C_w}$ can be synthesised and a general condition for this to occur is not known. We only report for its simplicity the following result based on the *arcsine law* [126, 127], noting that the most general approach to this synthesis is the solution of a computationally-intensive discrete optimisation problem [125].

Proposition 2.5 *(Synthesis of an aRBE by the arcsine law).* An aRBE $\mathbf{W}$ with row correlation matrix $\mathbf{C_w} : \forall j \in \{0, \ldots, n-1\}, (\mathbf{C_w})_{j,j} = 1$

may be synthesised if

$$\mathbf{C_t} = \sin\left(\frac{\pi}{2}\mathbf{C_w}\right) \succeq 0 \qquad (2.21)$$

by letting $\mathbf{T} \in \mathbb{R}^{m \times n}$ be an aRGE with row correlation matrix $\mathbf{C_t}$ and[a] $\mathbf{W} = \mathrm{sign}\,(\mathbf{T})$.

The *arcsine law* approach is summarised in the scheme in Fig. 2.3b. As a final remark, the requirement $\{\mathbf{C_w} \succeq 0 : \forall j \in \{0, \ldots, m-1\}, (\mathbf{C_w})_{j,j} = 1\}$ can be turned into an additional constraint[3] for Problem 2.1, specialising it to the design of correlation matrices that comply with at least one of the conditions of Proposition 2.5. However, this invalidates both (2.15) and the eigenvalue formulation (*i.e.*, the problem cannot be separated as in (2.14)). The resulting problem will become a semidefinite programming one, and will require a suitable solver: this approach is outside the scope of this thesis and will be pursued in a future communication.

2.3.3 Some Intuition on Maximum Energy Random Sensing Matrices and the RIP

The matter of computing the RIP for maximum energy sensing matrices $\mathbf{W}$ is non-trivial, even in the particular but significant case of an aRGE; in fact, most existing techniques for proving the RIP [17, 65] rely on the fact that $\mathbf{W}$ is drawn from an isotropic RME, and a modification of these arguments to account for the anisotropic case is not yet available to the best of the author's knowledge. What can be said is that, when $\mathbf{C_w} \succeq 0$ is the row correlation matrix of an aRGE $\mathbf{W} = \mathbf{T}\mathbf{C_w^{\frac{1}{2}}}$ with $\mathbf{T}$ drawn from a RGE (*e.g.*, as output from (2.15) with $\tau < \tau^*$), we have that by matrix similarity $\lambda_j(\mathbf{C_w^{\frac{1}{2}}}\mathbf{T^*T}\mathbf{C_w^{\frac{1}{2}}}) = \lambda_j(\mathbf{T^*T}\mathbf{C_w})$ and $\forall j \in \{0, \ldots, n-1\}$, $\lambda_{\min}(\mathbf{C_w})\lambda_j(\mathbf{T^*T}) \leq \lambda_j(\mathbf{C_w}\mathbf{T^*T}) \leq \lambda_{\max}(\mathbf{C_w})\lambda_j(\mathbf{T^*T})$ by application of [128, Theorem 8.12] to the symmetric PSD matrices $\mathbf{T^*T}$ and $\mathbf{C_w}$. This shows how the eigenvalues of $\mathbf{C_w}$ alter the distribution of those of $\mathbf{T^*T}$ by modifying their spread.

---

[a] Hence the name arcsine law, since $\mathbf{C_w} = \frac{2}{\pi}\arcsin(\mathbf{C_t})$.
[3] In that case, the energy constraint will be $\mathcal{E}_\mathbf{w} = n$ instead of $\mathcal{E}_\mathbf{w} = 1$.

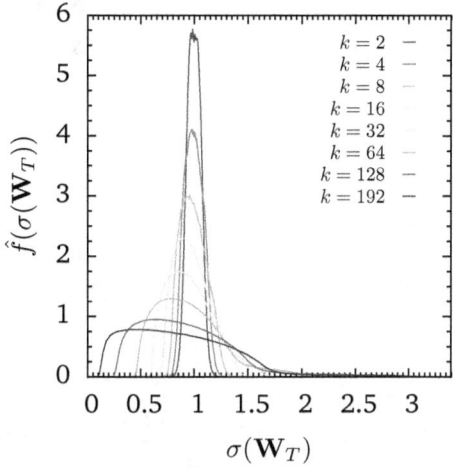

(a) The empirical PDF of singular values.

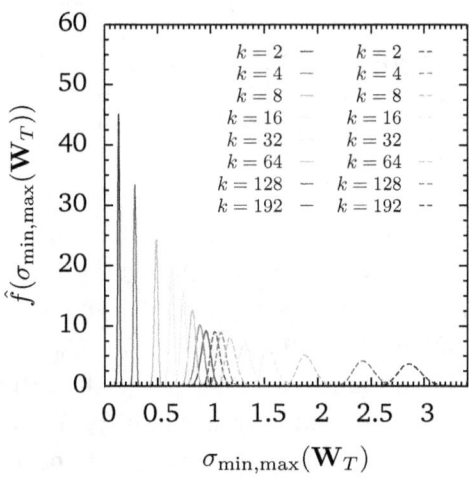

(b) Extreme singular values' PDF: $\sigma_{\min}(\mathbf{W}_T)$ (solid), $\sigma_{\min}(\mathbf{W}_T)$ (dashed).

Figure 2.4: Empirical distribution of the singular values of an exemplary aRGE with $\mathbf{C_w} \succeq 0$ corresponding to a k-column submatrix, with $m = 2^8$ as k varies.

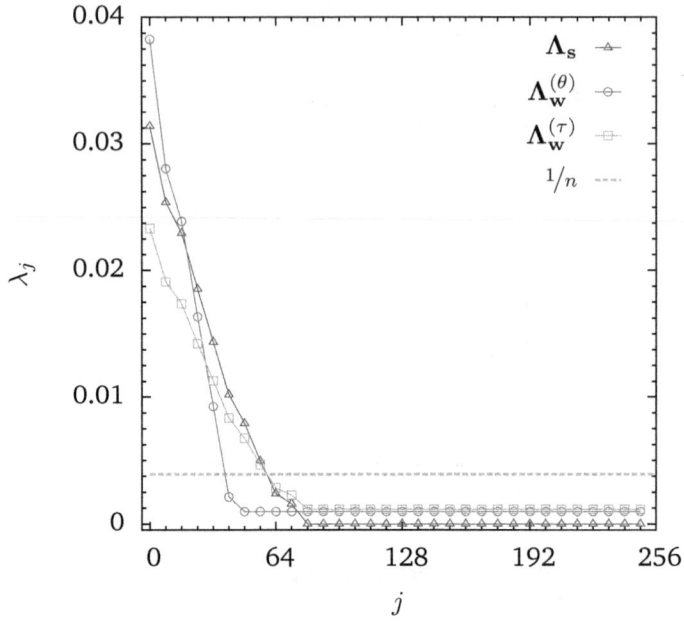

Figure 2.5: A comparison of the sequence of $n = 256$ eigenvalues of an exemplary $\mathbf{C_s}$ (triangles) with those output from the solution of Problem 2.1 with $\tau = 1/2$ (squares) and Problem 2.3 with $\theta = 1/4$ (circles).

The same effect can be seen when computing the singular values of $\mathbf{W}_T$ as in Fig. 2.4, where $\mathbf{C_w}$ was generated with an exponentially decreasing eigenvalue sequence between $\lambda_{\min}(\mathbf{C_w}) = 0.8957$ and $\lambda_{\max}(\mathbf{C_w}) = 4.1286$, attaining $\mathcal{L}(\mathbf{w}) = 0.0031$. The extreme singular values reported in Fig. 2.4b differ significantly from those in Fig. 1.7, i.e., they have a larger gap due to effects of the above observation. We can therefore expect an increase in the k-RIC depending on the eigenvalues of $\mathbf{C_w}$.

This is only reasonable, as the implicit information that the RIP would neglect is that, when $\mathbf{W}$ is drawn from an aRGE designed with the criteria discussed in this Chapter, the test vectors for a restricted isometry should not only be sparse but also suitably localised with the same $\mathbf{C_s}$ by which $\mathbf{W}$ was designed; and even so, one could specifically select a k-sparse instance of s aligned to the eigenvector of $\mathbf{C_w}$ corresponding to $\lambda_{\min}(\mathbf{C_w})$ and see that the amount of energy collected by y from such a vector is typically lower than the white case. This indicates that the RIP is unsuitable for evaluating the improvement of a maximum energy sensing matrix designed for a specific signal class having sparse and localised representations.

While a rigorous theory linking these two aspects is not provided here, we propose a variation of Problem 2.1 so that it verifies a simple constraint on the minimum eigenvalue of $\mathbf{C_w}$ as follows.

Problem 2.3 (*Maximum Energy Sensing Matrix Design (with minimum allocation constraint)*). Let $\mathbf{s}, \mathbf{w} \in \mathbb{R}^n$ be independent R.V.s with correlation matrices $\mathbf{C_s} \stackrel{\text{sed}}{=} \mathbf{U_s \Lambda_s U_s^*}$ and $\mathbf{C_w} \stackrel{\text{sed}}{=} \mathbf{U_w \Lambda_w U_w^*}$ respectively. A design of w that maximises (2.3) is obtained by letting $\mathbf{C_w^{(\theta)}} \stackrel{\text{sed}}{=} \mathbf{U_s \Lambda_w^{(\theta)} U_s^*}$, where

$$\Lambda_\mathbf{w}^{(\theta)} = \underset{\Lambda_\mathbf{w} \succeq 0}{\operatorname{argmax}} \operatorname{tr}(\Lambda_\mathbf{w} \Lambda_\mathbf{s})$$

$$\text{s.t.}$$

$$\mathcal{E}_\mathbf{w} = 1$$

$$\mathcal{L}(\mathbf{w}) \leq \mathcal{L}(\mathbf{s})$$

$$\Lambda_\mathbf{w} \geq \frac{\theta}{n} \mathbf{I}_n \quad (2.22)$$

with minimum allocation parameter $\theta \in (0,1)$.

Thus, θ in (2.22) ensures that a minimum amount of energy is allocated to each of the subspaces spanned by the eigenvectors in $\mathbf{U_s}$. The difference in the result of this choice and Problem 2.1 can be appreciated in the example of Fig. 2.5, as, *e.g.*, solved by CVX. A performance comparison between Problem 2.3 and 2.1 in critical cases where s has a different correlation matrix w.r.t. that used in solving the above problems is left for a future case study.

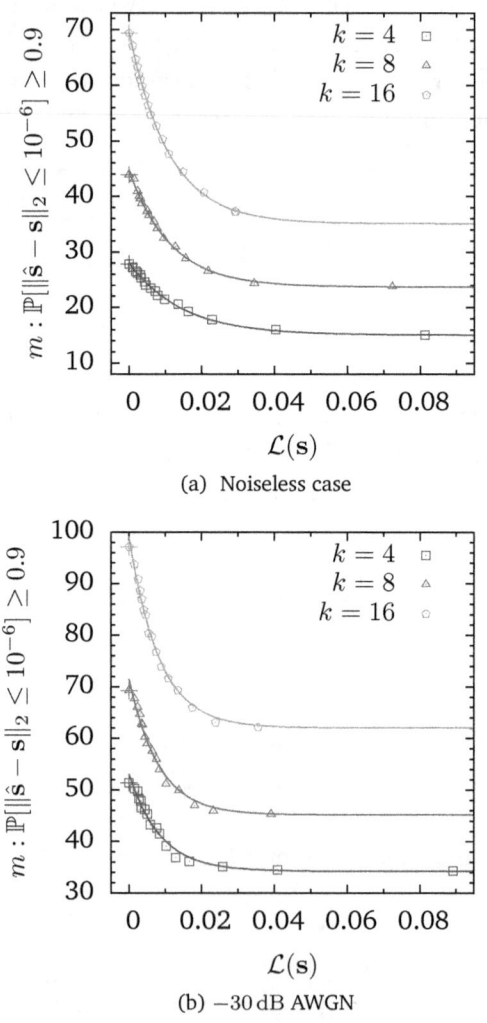

Figure 2.6: The minimum number of measurements needed by maximum energy sensing matrix design to attain probability 0.9 of successful reconstruction as a function of $\mathcal{L}(\mathbf{s})$.

2.4 Performance Evaluation

To demonstrate the effectiveness of the approach discussed in this Chapter, we now test some maximum energy sensing matrices $\mathbf{W}$ generated by Problem 2.1 on synthetic k-sparse and localised signals. We modulate the localisation of these signals by drawing the indices in each support $T, |T| = k$ with a non-equal assigned marginal probability: in our case, this probability assignment is taken with a doubly-triangular profile with maxima in $n/4 - 1$ and $3n/4 - 1$ and varying widths for the corresponding triangles; then we take $\mathbf{s}_T \sim \mathcal{N}(\mathbf{0}_k, \frac{1}{k}\mathbf{I}_k)$, thus producing different values for $\mathcal{L}(\mathbf{s})$ depending on the probability assignment. This choice allows the generation of k-sparse test signals with a variable amount of localisation. For each value of k and $\mathcal{L}(\mathbf{s})$, we generate 2000 instances of $\mathbf{s}$ and calculate the sample correlation $\widehat{\mathbf{C}}_\mathbf{s}$; this information is input to the scheme in Fig. 2.3a. By letting[4] $\tau = 1/2$ in the solution of Problem 2.1 we obtain $\mathbf{W}$ drawn from an aRGE that maximises the measurements' average energy. Then we evaluate the probability of successful recovery from $\mathbf{y} = \mathbf{W}\mathbf{s}$, *i.e.*, by BP; performing the same operation with a RGE (with i.i.d. entries, as obtained when $\tau = 0$ in Problem 2.1) allows us to draw a comparison between the two ensembles on the chosen sparse and localised signal class.

Fig. 2.6 plots the minimum number of measurements needed to attain probability 0.9 that the generated k-sparse instances are successfully recovered as a function of $\mathcal{L}(\mathbf{s})$ when a maximum energy $\mathbf{W}$ is used. Note that a classical RGE sensing matrix here corresponds to $\mathcal{L}(\mathbf{s}) = 0$ and needs a minimum number of measurements for successful reconstruction corresponding to the value of the curves at that abscissa. The results are specifically reported:

▶ in the noiseless case, *i.e.*, acquiring $\mathbf{y} = \mathbf{W}\mathbf{s}$ and decoding it by BP as solved by CPLEX [94], setting a recovery SNR requirement of $\text{SNR}_{\hat{\mathbf{s}}, \mathbf{s}} \geq 120\,\text{dB}$. This yields in the performances of Fig. 2.6a;

---

[4]We have found this mid-range choice to be suitable to observe signal recovery improvements without requiring any particular refinement of τ.

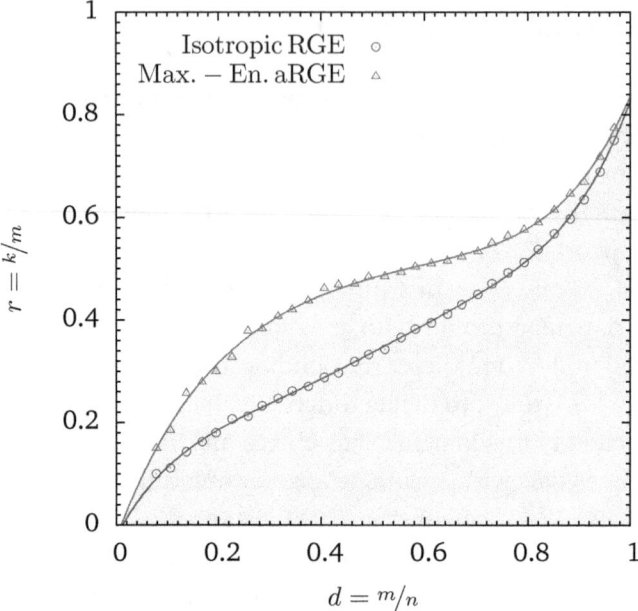

Figure 2.7: Empirical phase transition boundaries of BP for the anisotropic, maximum energy aRGE (triangles) *versus* the RGE (circles) as applied to a R.V. with $n = 256$, $\mathcal{L}(\mathbf{s}) = 0.03$. The filled regions indicate $\mathbb{P}[\hat{\mathbf{s}} = \mathbf{s}] \simeq 1$ for the two ensembles.

▶ when the input signal is perturbed as $\mathbf{y} = \mathbf{W}\mathbf{s} + \boldsymbol{\nu}$, with $\boldsymbol{\nu}$ *Additive White Gaussian Noise (AWGN)* setting a measurement SNR of 30 dB. In this case, signal recovery is carried out by BPDN as solved by CPLEX and considered successful when the recovery SNR is above a much lower requirement of $\text{SNR}_{\hat{\mathbf{s}},\mathbf{s}} \geq 20$ dB. This yields the performances of Fig. 2.6b.

Note how in both cases even relatively small localisation values (in agreement with those reported in Table 2.1) can be exploited to substantially decrease the number of measurements needed for successful recovery of the signal, and that the benefit increases as k increases.

As a final measure of performance and an at least partial recovery guarantee when the acquired signal complies with $\mathbf{C_s}$ used to solve

Problem 2.1, Fig. 2.7 reports a comparison of the phase transition boundary resulting from a maximum energy sensing matrix design (synthesised by an aRGE) against that attained by the RGE; the former design is here obtained when $\mathcal{L}(\mathbf{s}) = 0.03$ for s of dimensionality $n = 256$ and various levels of k. The empirical phase transition boundaries are reported in the same phase space described in Section 1.2.4 and correspond to cases of (m, n, k) in which 90% of the reconstructions of a noiseless signal (under the same conditions of Fig. 2.6a) are successful. Note how the maximum energy aRGE largely dominates the RGE in a fair comparison w.r.t. a sparse and localised signal, since the same number of measurements m allows the recovery of signals with a significantly larger k.

Summary

- The concept of localisation was introduced as an index to assess the deviation of a R.V. from the white case. Such a property can be exploited to adapt the design of sensing matrices when the signal being acquired follows this model, *i.e.*, the more a signal ensemble is localised, the more information can be extracted by means of a suitably adapted sensing matrix design.

- Sparse and localised R.V. are not trivially generated, yet are ubiquitous in most applications as the distribution of sparse representations is rarely isotropic for signal ensembles of interest. We here presented a generative model as a simulation method for such R.V.s; it is only a simplification w.r.t. the significantly more complex concept of PDFs for compressible and localised signals, that would provide a better model for natural signals.

- The proxy for "information extraction" adopted in the proposed adaptive sensing matrix designs is rakeness, *i.e.*, the average energy extracted from a R.V. representing a signal by another independent R.V. from which the rows of the sensing matrix are drawn. We provided numerical evidence that maximising the measurements' energy improves the probability of successful reconstruction of BP, as well as some intuition on the fact that doing so in expectation may improve the mapping of R.V.s with suitable localisation in their domain.

- Maximum energy sensing matrix design was reviewed as introduced by prior contributions in the context of this research; the solution of the related optimisation problem yields a correlation matrix according to which a sensing operator should be designed.

    Variations on the existing framework [29] were presented with the aim of stating the optimisation problems in terms of the formal definition of localisation. The synthesis procedures for aRGE and aRBE sensing matrices, as adapted to the target correlation matrix, were also reviewed to provide a complete view of the design procedure.

- A relationship with the RIP was numerically illustrated by highlighting how the extreme singular values of a RGE are modified by the non-white correlation matrix of an aRGE. A rigorous proof of the RIP would yield higher k-RIC values due to sparse R.V. in disagreement with the additional hypothesis of localisation; thus, a validation by the Donoho-Tanner phase transition is recommended to explain the improvement seen in the numerical evidence presented in this Chapter.

- The link between rakeness-based sensing matrix design and localisation was quantified by extensive simulations, closing a design flow for CS with maximum energy aRGE and aRBE sensing matrices. When the sparse representation of a signal is localised and such matrices are used to form the measurements, the empirical Donoho-Tanner phase transition w.r.t. BP was shown to improve w.r.t. the RGE.

LOW-COMPLEXITY DIGITAL SIGNAL COMPRESSION BY COMPRESSED SENSING

WIRELESS sensor networks [129] represent a relatively recent paradigm in information technology that poses some significant design challenges; each sensor node in a network must operate on a tight resource budget, the most limiting constraint being low power consumption in the acquisition, encoding and transmission of acquired data. To understand what terms concur in the resource budget of a sensor node, we report the traditional scheme of Fig. 3.1: a sensor node digitises an analog signal by means of Nyquist-rate *Analog-to-Digital (A/D)* conversion; the acquired samples are compressed on-board by a lossy and/or lossless encoder that most often operates in a suitable transform domain (*e.g.*, derived from a DCT or DWT), quantising the transformed coefficients and performing lossless encoding to eliminate residual redundancy in the encoded bitstream; the result is then transmitted to a remote location or stored on a suitable local memory. The received (or stored) data will then be processed by an off-board processing node to decode the information content up to the data fidelity allowed by measurement noise and lossy encoding.

Since power consumption in a sensor node is generally dominated by data transmission, minimising its rate by suitable lossy or lossless encoding stages is critical in administering the nodes' resources.

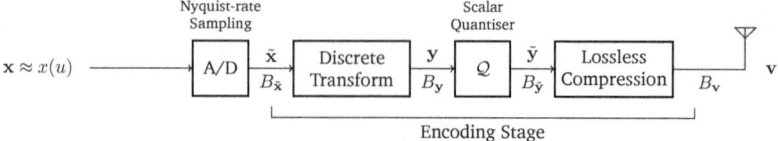

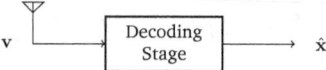

(a) Sensor node; the bitstream lengths are denoted by $B_{..}$.

(b) Processing node; the decoding stage contains the inverse operations of the encoding stage, when invertible.

Figure 3.1: A standard sensor node – processing node pair, highlighting the role of digital signal compression prior to transmission. Channel coding is regarded as part of the transmitter/receiver.

We here assume that signal acquisition and data compression are performed by low-power, low-complexity sensor nodes that transmit their encoded bitstreams to a processing node providing much larger computational power. Such an extreme resource asymmetry limits the use of multimedia compression schemes designed on the opposite assumption that the encoding is performed only once (therefore as computationally demanding as required) whereas decoding is performed multiple times as users access the information content (therefore as lightweight as possible).

Compression schemes that exploit only a few very elementary computations, *e.g.*, easily implemented in low-complexity fixed-point digital architectures are appealing to cope with such resource constraints. In this view, CS could be seen as a lossy compression scheme whose encoding stage is obtained by simple projection of the signal onto a small number of antipodal-valued vectors, *i.e.*, the rows of the RBE or aRBE discussed in Section 1.2.3; thus, its computational and digital hardware complexity is expected to be minimum. On the other hand and in an asymmetric fashion, its decoding stage would require the computational effort of solving some convex optimisation problems mentioned in Section 1.3 (with the alternative of using greedy algorithms, *e.g.*, as reviewed in [102]).

Existing investigations that analyse CS as a digital-to-digital lossy compression (see [25]) show that its rate-distortion performances [130] are asymptotically sub-optimal w.r.t. common transform-coding techniques. Although correct, these analyses do not account for the digital hardware requirements of such transforms, that often require floating-point multiplications. This would make CS with the RBE (or aRBE) an extremely lightweight multiplierless option to standard compression schemes. In addition, CS may be paired with further compression stages such as simple lossless Huffman coding to attain lower code rates.

Thus, the task of encoding a signal by CS is well-suited to the tight resource requirements of sensor nodes, whereas signal recovery will be carried out by a central node receiving the encoded bitstreams. By making this analogy between the asymmetry in the requirements of sensor networks and CS, we here explore the possibility of using the latter as a digital signal compression scheme whose complexity is well-matched by its effectiveness as a lossy compression.

As a practical illustration of this application, we compare the performances of CS with some reference compression schemes as applied to the particular, yet significant case of ECG signal compression; in addition, we show that a direct application of the principles developed in Chapter 2 allows a further, significant code rate reduction for the proposed compression scheme.

3.1 Lossy Compression Schemes for Biosignals

We here consider the specific case of ECG signals as a relevant example for the development of wireless sensors. The appeal of such signals is due to the fact that they exhibit a quasi-stationary behaviour over time, as they convey information on an essentially periodic phenomenon. Thus, n-samples windows x (*i.e.*, when they are considered as a R.V.) of this signal class are typically compressible w.r.t. a suitable DWT with a sparsity level k having only minor fluctuations.

The standard approach to acquiring such signals is depicted in Fig. 3.1: the analog ECG is first acquired by A/D conversion that

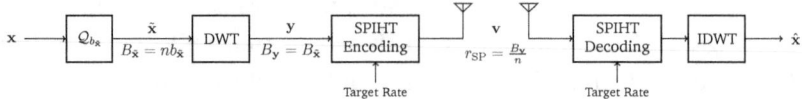

(a) Huffman coding: encoding and decoding stages

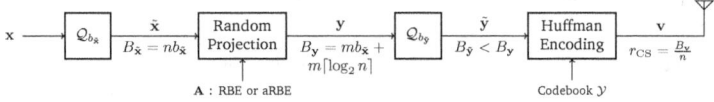

(b) Set partition coding: encoding and decoding stages

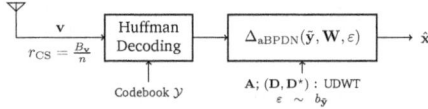

(c) Compressed sensing: encoding stage

(d) Compressed sensing: decoding stage

Figure 3.2: Block diagram of the three evaluated compression schemes. Channel coding is regarded as part of the transmitter/receiver.

discretises it into n Nyquist-rate samples collected in $\mathbf{x}$. Moreover, the A/D converter embeds a quantisation of the signal range into the *Pulse Code-Modulated (PCM)* samples $\tilde{\mathbf{x}} = \mathcal{Q}_{b_{\tilde{\mathbf{x}}}}(\mathbf{x})$, with $\mathcal{Q}_{b_{\tilde{\mathbf{x}}}}$ denoting uniform[1] scalar quantisation with $b_{\tilde{\mathbf{x}}}$ bits per sample (bps) tuned to fit the full signal range. The task of encoding $\mathbf{x}$ prior to transmission can be divided in two stages (the bitstream lengths are denoted by $B.$):

1. a lossy encoding stage that allows for a reduced-size bitstream $\tilde{\mathbf{y}}$ by accepting some information loss w.r.t. $\tilde{\mathbf{x}}$. This is divided in a discrete transform that maps $\tilde{\mathbf{x}}$ in a domain where a compressible behaviour is observed (as in Definition 1.4) followed by an

---

[1]The integration of non-uniform, minimum-distortion quantisers at the A/D converter is generally a technologically complex task; for this reason, we limit this study to uniform quantisers.

additional quantisation step, where information loss is allowed with the purpose of reducing the code rate;

2. a lossless encoding stage that eliminates the remaining redundancy in $\tilde{y}$ by operating on its bitstream, returning a compressed binary string v at the output. A typical example of such a stage is any entropy coding scheme [131].

The two stages achieve for an n-samples window a code rate of $r = B_v/n$ bps with a total of B_v bits in the encoded bitstream. In particular, we here evaluate the possibility of using CS as digital signal compression scheme which applies a dimensionality reduction on $\tilde{x}$, that is suitably used as a discrete transform in the scheme of Fig. 3.1. We now proceed as follows: firstly, we introduce two common compression techniques (one lossless, and one lossy w.r.t. $\tilde{x}$) that may be considered as terms of comparison for this task; then we discuss a lossy compression scheme based on CS and tune it to attain optimal performances; finally, we compare the three techniques as optimally tuned.

3.1.1 Huffman Coding

A low-complexity lossless compression scheme considered for this comparison amounts to processing the PCM samples in $\tilde{x}$ with standard *Huffman Coding (HC)* [131]; this entropy coding technique takes a binary string as an input, and encodes it by a prefix-free variable-length code. This code is based on the construction of an optimal codebook based on the probability distribution of the input, *i.e.*, the most probable symbol in the input string is encoded by the shortest codeword, and so forth in the construction of a binary tree that uniquely encodes all non-zero probability symbols.

The codebook is here assumed to be known *a priori* and is practically trained on the empirical distribution of a very large set of PCM samples (in particular, of a large dataset of ECG samples). Since this training set might not contain all possible words an *escape* codeword is added to the codebook, followed by $\lceil \log_2 q \rceil$ bits to represent all of the q symbols not appearing in the above set. Thus, the "quality loss" here is only due to the inevitable uniform quantisation of x into x̃ caused by A/D conversion.

This compression scheme requires a minimum amount of computational resources: after the signal is quantised, we straightforwardly encode x̃ by using a lookup table that maps its fixed-length words to variable-length codewords in the encoded bitstream v. Thus, if sufficient memory is available at the sensor node to store the optimal codebook, HC achieves a code rate r_{HC} with no fixed-point signal processing operation involved. This scheme is depicted in Fig. 3.2a.

3.1.2 Set Partition Coding of Wavelet Coefficients

To the other end of our comparison, we consider the application of *Set Partitioning In Hierarchical Trees (SPIHT)* [132] that serves as a basic building block for many wavelet-domain digital signal compression schemes (see Fig. 3.2b). The SPIHT encoder operates on the DWT coefficients of x̃ (in particular, the authors of [132] recommend 9/7 biorthogonal wavelets [52]) by constructing a map of their significance w.r.t. their magnitudes and dependencies in a tree representation of the wavelet coefficients.

The critical arithmetic complexity in this lossy encoding is in implementing the chosen DWT that, as efficient [133] and specific [134] as can be made, requires fixed-point multiplications with quantised filter coefficients. Such a complexity is considered high for straightforward integration into low-resources digital processing stages for sensor nodes; we will report its attained code rates r_{SP} as a reference case that is expected to outperform the other schemes, with the purpose of showing how CS is capable of achieving acceptable rates with a low-complexity multiplierless transform.

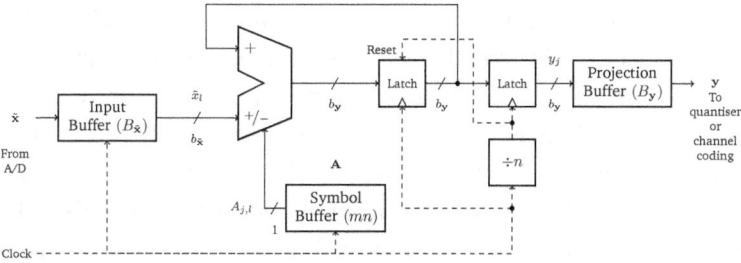

Figure 3.3: A digital, multiplierless hardware implementation of the CS encoding stage with RBE matrices, using a single accumulator and fixed-point arithmetic; the buffers are local registers of size denoted by $(\cdot)$ bit; the dashed lines denote synchronisation signals.

3.1.3 Lossy Compression by Compressed Sensing

The Encoding Stage

As mentioned in Section 3.1.3 a dimensionality reduction is simply obtained as $\mathbf{y} = \mathbf{A}\mathbf{x}$; in this Chapter, we will refer to $\mathbf{A}$ as the *encoding matrix* to emphasise that it is implemented in a digital-to-digital fashion; in particular, we let $\mathbf{A} \in \{-1, +1\}^{m \times n}, m < n$ since we want to implement it in very low-complexity digital hardware.

The proposed encoding stage is reported in Fig. 3.2c and summarised as follows. As dimensionality reduction is here performed in the digital domain, we will operate on quantised $\tilde{\mathbf{x}}$; thus, the encoding operation is actually $\mathbf{y} = \mathbf{A}\tilde{\mathbf{x}}$ represented by m digital words. Their wordlength will be at most $b_{\mathbf{y}} = b_{\tilde{\mathbf{x}}} + \lceil \log_2 n \rceil$ bits since each y_j is obtained by an inner product of the PCM samples in $\tilde{\mathbf{x}}$ with a vector of sign changes, *i.e.*, $y_j = \pm \tilde{x}_0 \pm \tilde{x}_1 \pm \ldots \pm \tilde{x}_{n-1}$. This operation can be conveniently mapped on mn cycles of a single accumulator, *i.e.*, by an extremely simple multiplierless fixed-point digital scheme (see Fig. 3.3).

To reduce the rate of the encoded bitstream, we quantise $\mathbf{y}$ by a second uniform scalar quantiser as $\tilde{\mathbf{y}} = \mathcal{Q}_{b_{\tilde{\mathbf{y}}}}(\mathbf{A}\tilde{\mathbf{x}})$ with $b_{\tilde{\mathbf{y}}} \leq b_{\mathbf{y}}$; $\mathcal{Q}_{b_{\tilde{\mathbf{y}}}}$ is scaled to operate in the range of $\mathbf{y}$ but keeps only $b_{\tilde{\mathbf{y}}}$ *Most Significant Bits (MSBs)* from each y_j. We also note that the alternative of a non-uniform, minimum-distortion scalar quantiser (*i.e.*, a Lloyd-Max quantiser [135]) could indeed be pursued here as only requiring the implementation of a suitable pre-distortion prior to uniform quantisation, whereas vector quantisation commonly requires more computational effort on the encoder [34]. In a low-complexity perspective we assume that a uniform quantiser is the simplest choice for this task, although this alternative is indeed worth exploring (and has already been addressed in some works [136, 137]).

To further compress the encoded bitstream we also evaluate the option of applying lossless HC with an optimal codebook trained on the empirical PMFs of each element of $\tilde{\mathbf{y}}$, that are approximately Gaussian-distributed due to the mixing effect of $\mathbf{A}$ (this assumption will be further cleared out later in Section 7.2). Thus, the encoded bitstream $\mathbf{v}$ attains a code rate r_{CS} that depends on $(m, b_{\tilde{\mathbf{x}}}, b_{\tilde{\mathbf{y}}})$, the choice of $\mathbf{A}$ and the presence or absence of HC.

Maximum Energy Encoding Matrices

We now proceed to discussing a further degree of freedom in the choice of $\mathbf{A}$ as drawn from a suitably chosen maximum energy aRBE rather than the classic choice of an RBE. Although assuming $\mathbf{A}$ as drawn from the RBE fits equally well any kind of signal [17], we have shown how localisation can be leveraged to design $\mathbf{A}$ from anisotropic RMEs to maximise the average energy of $\mathbf{y}$. As explored in Fig. 2.7, this rakeness-based approach to compressed sensing was empirically shown to lower the requirements on the minimum m to attain successful signal recovery; we therefore use it as another encoder-side option to reduce the code rate.

To carry out this adaptation, we recall that the quasi-stationary behaviour of ECGs allows for a meaningful estimation of the signal's correlation matrix $\mathbf{C_x} \stackrel{\text{sed}}{=} \mathbf{U_x \Lambda_x U_x^*}$, where $\mathbf{U_x}$ may be used to perform optimal transform-coding by the KLT [138]. In many applications, the stationarity of $\mathbf{C_x}$ over time is insufficient, and the update and transmission of its estimate $\widehat{\mathbf{C}}_{\mathbf{x}}$ makes the KLT usually disadvantageous w.r.t. computing other transforms. However, for this particular type of

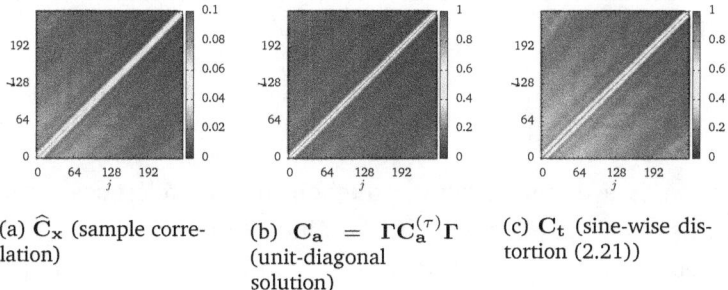

(a) $\widehat{\mathbf{C}}_\mathbf{x}$ (sample correlation)

(b) $\mathbf{C_a} = \mathbf{\Gamma C_a^{(\tau)} \Gamma}$ (unit-diagonal solution)

(c) $\mathbf{C_t}$ (sine-wise distortion (2.21))

Figure 3.4: Correlation matrices related to a maximum energy aRBE encoding matrix design.

signal the estimated $\mathbf{C_x}$ is not only stable, but typically attains high values of $\mathcal{L}(\mathbf{x})$ in (2.2).

To do so, given the quasi-stationary behaviour of ECGs, we here apply the synthesis scheme of Fig. 2.3b initialising it with the sample correlation matrix $\widehat{\mathbf{C}}_\mathbf{x}$ depicted in Fig. 3.4a as estimated from a large training set of 10^4 instances of $\mathbf{x}$. The synthesis problem is solved with the purpose of obtaining an aRBE from which $\mathbf{A}$ can be drawn: by (2.15) we obtain $\mathbf{C_a^{(\tau)}}$ for $\tau = 1/2$, then we proceed to fulfilling the synthesis condition of Proposition 2.5 (i.e., the input matrix should have only ones on its diagonal) by scaling the matrix $\mathbf{C_a^{(\tau)}}$ with $\gamma = \mathrm{diag}\left(\mathbf{C_a^{(\tau)}}\right)^{-1}$ (i.e., the element-wise inverse of the diagonal of the latter correlation matrix), $\mathbf{\Gamma} = \mathrm{diag}\left(\gamma^{\frac{1}{2}}\right)$, letting $\mathbf{C_a} = \mathbf{\Gamma C_a^{(\tau)} \Gamma}$; this intermediate matrix is reported in Fig. 3.4b. The scaling allows us to obtain $\mathbf{C_t}$ by plugging $\mathbf{C_a}$ in (2.21) and is pictorially reported in Fig. 3.4c; the final verification that $\mathbf{C_t} \succeq 0$ allows us to conclude that $\mathbf{A}$ can indeed be generated by an aRBE using the scheme at the bottom of Fig. 2.3b. In a sense, this synthesis strategy can be considered similar to a KLT with antipodal-valued random projection vectors, yet more robust due to how Problem 2.1 is solved with a localisation constraint.

Thus, the resulting maximum energy $\mathbf{y}$ will have by design (i.e., by the very definition of the optimisation criterion (2.4)) a larger variance than that produced by the classic RBE case, so the following quantiser and Huffman code in Fig. 3.2c will require an adaptation to the new distribution of **y**.

The Decoding Stage

Since **A** is a dimensionality reduction and **y** undergoes a second quantisation, this scheme is by definition lossy. However, we have previously recalled some theoretical guarantees that relate the sparsity of **x** w.r.t. **D** and the minimum number of measurements $\overline{m} = \mathcal{O}(k \log(p/k))$ ensuring that **x** may be stably recovered from $\tilde{\mathbf{y}}$ even in the presence of quantisation noise. This guarantee allows us to consider the possibility that, when **x** is sufficiently sparse w.r.t. **D**, some denoising may indeed be possible by a suitable choice of dictionary and recovery algorithm.

The decoding stage discussed in this Section is reported in Fig. 3.2d. As a recovery algorithm we have considered *Analysis BPDN (aBPDN)*, i.e., (1.24) with $\varepsilon \geq 0$ essentially set in excess of the quantisation noise variance introduced by $\mathcal{Q}_{b_{\tilde{x}}}$ and $\mathcal{Q}_{b_{\tilde{y}}}$ on **x**.

As for $(\mathbf{D}, \mathbf{D}^\star)$ we assume they are the synthesis and analysis operators of an *Undecimated DWT (UDWT)* that is obtained from an orthonormal DWT by essentially removing the decimation and upsampling operations in its filterbank. This arrangement of a signal recovery algorithm and analysis-sparsity was shown to be robust w.r.t. additive noise in several contributions [49, 50, 139]; we aim at leveraging this robustness to mitigate the impact of quantisation on the quality of $\hat{\mathbf{x}}$.

3.2 Performance Evaluation

In this Section we evaluate the performances after decoding of the schemes in Fig. 3.2, with an emphasis on CS and its variants. We adopt the average SNR of the decoded signal, $\text{ASNR}_{\hat{\mathbf{x}}, \mathbf{x}}$ in (1.32) as a performance index, where $\hat{\mathbf{x}}$ is the corresponding decoded output of each of the considered techniques.

3.2.1 Signal Generation

We here use a synthetic ECG generator [140] to produce 10^4 training instances of **x** with $n = 256$, corresponding to 1 s windows sampled at 256 Hz. The parameters of the generator are randomly drawn to obtain a training set oscillating at various heart rates and not corrupted by intrinsic or quantisation noise. Each window is then quantised to its PCM samples $\tilde{\mathbf{x}}$ at $b_{\tilde{\mathbf{x}}}$ bits per sample. Since the ECG PCM samples generally have a high crest factor $\mathrm{CF} = 20 \log_{10} \frac{\sqrt{n}\|\mathbf{x}\|_\infty}{\|\mathbf{x}\|_2} \approx 11\,\mathrm{dB}$ they are non-uniformly distributed in the quantiser range. Thus, the SNR w.r.t. uniform white quantisation noise is estimated as $\mathrm{SNR}_{\mathcal{Q}_{b_{\tilde{\mathbf{x}}}}} = 10 \log_{10} \frac{\hat{\mathbb{E}}[\|\mathbf{x}\|_2^2]}{\hat{\mathbb{E}}[\|\tilde{\mathbf{x}}-\mathbf{x}\|_2^2]} \approx 6.02\, b_{\tilde{\mathbf{x}}} - 11\,\mathrm{dB}$ (as will be reported in Fig. 3.6, 3.7) where the second term is indeed due to the ECG signals' high crest factor.

3.2.2 Some Details on the Decoder

The choice of a suitable wavelet family for the UDWT and of a decoding algorithm for solving (1.24) are crucial for a fair evaluation of CS. We here assume that $(\mathbf{D}, \mathbf{D}^\star)$ are those of the Symmlet-6 UDWT with $J = 4$ sub-bands (i.e., $p = (J+1)n$) [52, Chapter 5.2], and adopt this transform for signal recovery. For what concerns aBPDN, we solve (1.24) by the UnLocBox [141] implementation of Douglas-Rachford splitting [113] with the data fidelity constraint of (1.24) tuned to the noise norm $\varepsilon = \|\tilde{\mathbf{y}} - \mathbf{A}\mathbf{x}\|_2$ and ensuring that the algorithm converges up to a relative variation of 10^{-7} in the objective function.

3.2.3 Measurements' Quantisation Effects

The main noise sources in the schemes considered here are the uniform PCM quantisers $\mathcal{Q}_{b_{\tilde{\mathbf{x}}}}, \mathcal{Q}_{b_{\tilde{\mathbf{y}}}}$. While the former is common to all evaluated schemes, the latter is only used in the CS encoding to reduce each element of **y** to $b_{\tilde{\mathbf{y}}} < b_{\mathbf{y}}$ bits. Since these measurements are approximately Gaussian-distributed (as will be discussed in Section 7.2) $b_{\mathbf{y}} = b_{\tilde{\mathbf{x}}} + \lceil \log_2 n \rceil$ might exceed the precision actually required to represent **y** with negligible losses. Thus, to explore the effect of $b_{\tilde{\mathbf{y}}}$

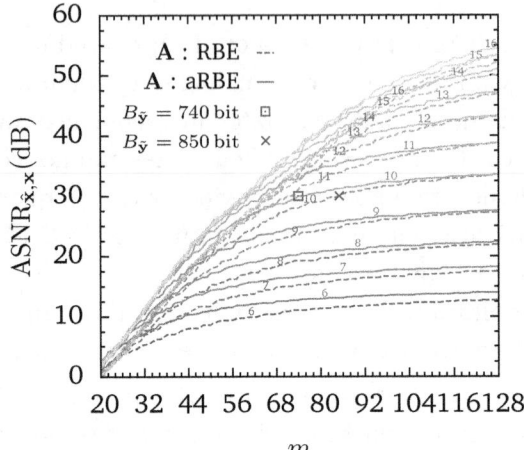

(a) Measurements $\tilde{y}$ quantised with $b_{\tilde{y}} = b_{\tilde{x}}$ bits (values of $b_{\tilde{y}}$ reported on curves)

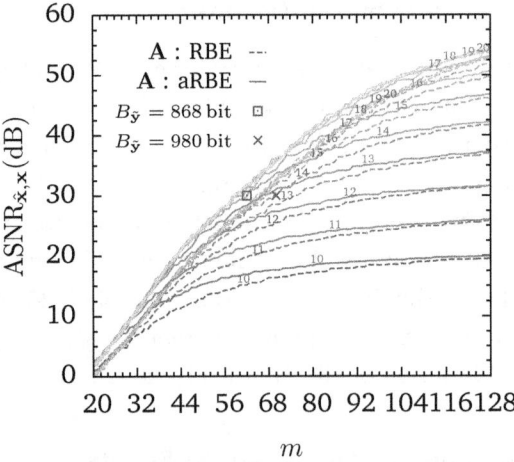

(b) Measurements $\tilde{y}$ quantised with $b_{\tilde{y}} = b_{\tilde{x}} + \lceil \frac{1}{2}\log_2 n \rceil$ bits (values of $b_{\tilde{y}}$ reported on curves)

Figure 3.5: $\mathrm{ASNR}_{\hat{x},x}$ for the i.i.d. RBE (dashed) and maximum energy aRBE (solid) CS with different quantisation policies. For both figures $b_{\tilde{x}} = 6, \ldots, 16$. For $b_{\tilde{x}} = 10$, the points corresponding to bit budgets that allow an $\mathrm{ASNR}_{\hat{x},x} \approx 30\,\mathrm{dB}$ highlight the i.i.d. RBE case (cross) and maximum energy aRBE case (square).

we (i) encode by CS the ECG training set and train $\mathcal{Q}_{b_{\tilde{y}}}$ with either $b_{\tilde{y}} = b_{\tilde{x}}$ or $b_{\tilde{y}} = b_{\tilde{x}} + \lceil \frac{1}{2} \log_2 n \rceil$ (ii) apply the same operation on 64 new test instances, solve (1.24) and compute $\text{ASNR}_{\hat{x},x}$ while varying $m = 20, \ldots, 128$ (up to $m = n/2$), $b_{\tilde{x}} = 6, \ldots, 16$. Moreover, we run the very same procedure for rakeness-based CS trained as in Section 3.2.2 and with a suitably scaled range for $\mathcal{Q}_{b_{\tilde{y}}}$ to compensate for the fact that the measurements have a larger average energy.

The results of this procedure are reported in Fig. 3.5, and allow us to observe that (i) rakeness-based CS with maximum energy aRBE outperforms standard CS with the i.i.d. RBE in all the examined cases; (ii) the quality gain obtained by using more bits for both ($b_{\tilde{x}}, b_{\tilde{y}}$) progressively saturates at an $\text{ASNR}_{\hat{x},x}$ limit imposed by the sparsity level of ECG signals; (iii) for a fixed value of $b_{\tilde{x}}$, the total bit budget $B_{\tilde{y}} = mb_{\tilde{y}}$ required to reach an $\text{ASNR}_{\hat{x},x}$ target hints at how redundant the chosen quantisation policy is. This quantity is highlighted in both Fig. 3.5a,3.5b, and shows how the quality improvement of choosing a more accurate quantiser $\mathcal{Q}_{b_{\tilde{y}}}$ for $\tilde{y}$ must be matched with a smaller m, and in particular that $b_{\tilde{y}} = b_{\tilde{x}}$ is a better choice for achieving lower code rates with CS.

3.2.4 Rate Performances

Given the observed quantisation effects, to understand which uniform scalar quantiser $\mathcal{Q}_{b_{\tilde{y}}}$ enables the lowest code rate, $\tilde{y}$ must be post-processed by optimally-trained HC. In addition, we here assess how this attained rate, r_{CS}, compares with the rate performances achieved by the other schemes (Fig. 3.2) at some fixed target decoding performances, i.e., $\text{ASNR}_{\hat{x},x} = \{25, 30, 35, 40, 45, 50\}$ dB. For a fair comparison, SPIHT for ECGs [132] is run from the authors' code by fitting instances of $\tilde{x}$ into full frames of 1024 PCM samples quantised at different $b_{\tilde{x}}$. The SPIHT encoder takes r_{SP} as an input, which we vary in $[1/n, 2]$; the minimum r_{SP} that guarantees the target $\text{ASNR}_{\hat{x},x}$ after decoding is then reported in Fig. 3.6, 3.7. As a further reference, we report the rates of uniform PCM quantisation and its optimal HC, achieving a rate r_{HC}; since it is lossless, achieving an $\text{ASNR}_{\hat{x},x}$ target

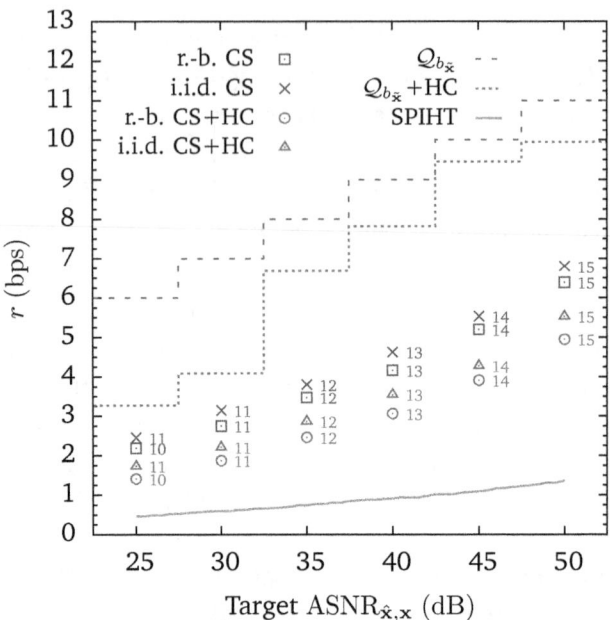

Figure 3.6: Achieved code rates of the evaluated compression schemes and their variants for the chosen $\text{ASNR}_{\hat{x},x}$ target specifications; the CS measurements $\tilde{y}$ are quantised with $b_{\tilde{y}} = b_{\tilde{x}}$; the value of $b_{\tilde{y}}$ that allows a given rate is reported to the right of each marker.

depends on $b_{\tilde{x}}$. While the average codeword length (and r_{HC}) could be estimated as the entropy of PCM samples, to account for the presence of escape symbols we run this encoding to find the actual r_{HC} of the test set.

These two reference methods are compared with various embodiments of CS (*i.e.*, with or without HC; with different quantisation policies; with or without a rakeness-based, maximum energy aRBE encoding matrix design) in Fig. 3.6, 3.7. It is observed that the rates attained in Fig. 3.6 are generally lower than those in Fig. 3.7, thus confirming the benefits of assuming $b_{\tilde{y}} = b_{\tilde{x}}$. In addition, (*i*) the use of HC on the measurements reduces significantly the code rate of CS, as also does the use of rakeness-based encoding matrices (*ii*) by considering r_{CS} of rakeness-based CS with HC, Fig. 3.6 shows that an

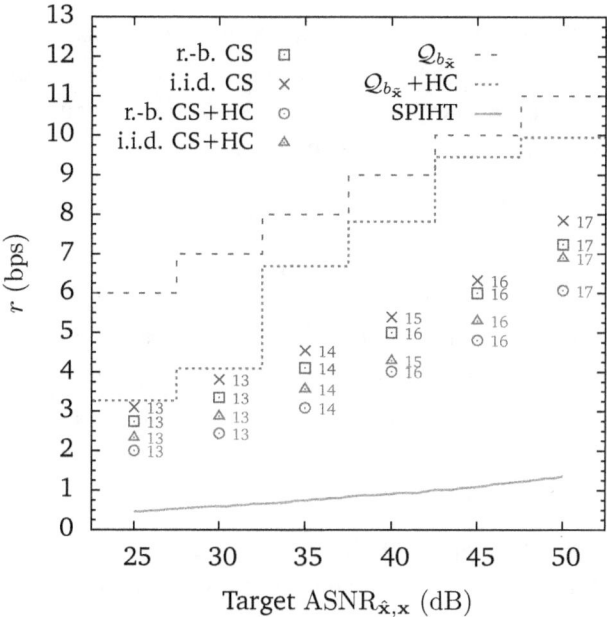

Figure 3.7: Achieved code rates of the evaluated compression schemes and their variants for the chosen $\mathrm{ASNR}_{\hat{\mathbf{x}},\mathbf{x}}$ target specification; the CS measurements $\tilde{\mathbf{y}}$ are quantised with $b_{\tilde{\mathbf{y}}} = b_{\tilde{\mathbf{x}}} + \lceil \frac{1}{2} \log_2 n \rceil$; the value of $b_{\tilde{\mathbf{y}}}$ that allows a given rate is reported to the right of each marker.

$\mathrm{ASNR}_{\hat{\mathbf{x}},\mathbf{x}} \approx 25\,\mathrm{dB}$ is achieved at $b_{\tilde{\mathbf{y}}} = b_{\tilde{\mathbf{x}}} = 10\,\mathrm{bit}$ by $r_{\mathrm{CS}} \approx 1.41\,\mathrm{bps}$, while $r_{\mathrm{HC}} = 3.27\,\mathrm{bps}$. At higher $\mathrm{ASNR}_{\hat{\mathbf{x}},\mathbf{x}}$ targets, CS is increasingly advantageous, placing itself at less than 50% of the code rate of PCM with optimal HC.

Thus, we conclude that as a lossy compression CS can achieve low code rates; at the same time, its computational complexity on the encoder side is extremely low and requires no multiplication. Given these low requirements, it lends itself as an agile lossy scheme for resource-constrained signal compression applications. This said, many degrees of freedom are still to be explored to improve upon these results; as mentioned, since the second scalar quantiser is fully digital and can be arbitrarily tuned, Lloyd-Max quantisation could be used to reduce the measurements' distortion for a given code rate, exploiting the fact that their statistics are well approximately Gaussian-distributed. In addition, we expect a the choice of a closer modelling of quantisation noise in the signal recovery problem [142] to enable even lower code rates for the chosen distortion specifications.

Summary

- ▶ CS may be used as a lossy digital signal compression algorithm, with the aim of reducing the transmission rate of a sensor node. The advantage of CS is having an extremely low-complexity digital implementation potentially requiring a single accumulator and no floating-point multiplication when an RBE or aRBE is used as a sensing matrix.

- ▶ Reducing the length of the encoded bitstream is possible by combining CS with HC and with aRBE maximum energy sensing matrices.

- ▶ The trade-off between the number of measurements collected by CS and their quantisation must be carefully administered, as both terms concur in the total bit budget of the encoded bitstream.

- ▶ A comparison with a classical signal compression algorithm such as SPIHT shows that the rates attained by CS are sub-optimal, yet the complexity reduction provided by the latter is still an appealing asset for the design of a compression scheme at a negligible cost.

4
MAXIMUM ENTROPY SENSING MATRIX DESIGNS FOR LOCALISED SIGNALS

We now investigate the design of random sensing matrices constructed from a deterministic, fixed set of vectors, *e.g.*, as the PHE or PFE discussed in Section 1.2.3. In continuity with Chapter 2 the property leveraged in this context is localisation; however, the proxy adopted to optimise this class of sensing matrices leverages the *maximum entropy* principle [143] with the aim of selecting an optimal set of m vectors based on the analysis of their covariance.

Since the exact solution of the resulting selection problem is NP-hard due to the nature of the maximum entropy sampling problem a lightweight heuristic algorithm is introduced to generate a pool of sensing matrices belonging to the PHE (or any similarly constructed ensemble) yielding m near-maximum entropy measurements. The criterion and its heuristic implementation allow us to devise a strategy to determine optimal sensing matrices in the PHEs; some results are provided when applying our method to CS of small images and ECGs. The improvements found in terms of signal recovery performances, despite the non-minimum coherence between the sensing matrix and the chosen sparsity bases, indicates a partial overcoming of the non-universality of the PHE.

4.1 Compressed Sensing with Deterministic Ensembles

We here focus on *deterministic ensembles, i.e.*, RMEs constructed from deterministic sets of vectors such as the PHE, that become of practical interest when the choice of sensing vectors is fixed or constrained by the acquisition mechanism – the most celebrated application related to this scenario is magnetic resonance imaging by CS [19], which uses the PFE.

Thus, we assume to be limited to an orthonormal *design space* of feasible sensing vectors $\{\mathbf{a}_j\}_{j=0}^{n-1}, \mathbf{a}_j \in \mathbb{R}^n$ collected in the rows of a full-rank $\mathbf{A}_{n \times n}$. The sensing matrix, which we now denote $\mathbf{A}^S$, is constructed by extracting m row vectors at random from $\mathbf{A}$ with indices in the subset $S = \{0, \ldots, m-1\}, |S| = m$. In absence of other assumptions, the $\binom{n}{m}$ possible $\mathbf{A}^S$ are considered equally good candidates in the corresponding RME. Throughout this Chapter, the notation $\cdot^S$ will indicate the selection of m-cardinality row submatrices or subsets S in a matrix or vector, and by $\cdot^\star$ optimal values.

Although coherence between the sensing matrix $\mathbf{A}$ and the dictionary $\mathbf{D}$ should be made as small as possible as recalled in Section 1.2, it is suggested [50] that it only needs to be bounded in many cases. Since the design space is considered fixed, *e.g.*, by the properties of the acquisition mechanism, and the sparsity basis is set by the nature of the signal being sampled, coherent pairs of $(\mathbf{A}, \mathbf{D})$ will occur in some prospective applications of CS.

In particular, we here emphasise the case of the Hadamard matrix $\mathbf{H}_n$ that forms the sensing vectors of the PHE, as introduced in Section 1.2.3. This transform is particularly suitable to both analog, optical and digital implementations. Due to its recursive structure it can also be fully computed by a divide-and-conquer algorithm [144] in $\mathcal{O}(n \log_2 n)$ instead of $\mathcal{O}(mn) = \mathcal{O}(nk \log_2 n)$ floating point operations. This appealing low-complexity property drives the idea of finding optimal PHE sensing matrices $\mathbf{A}^S$ for generic dictionaries $\mathbf{D}$ (here assumed to be ONBs for the sake of simplicity), noting that $\mathbf{A}^S$ will exhibit non-minimum coherence w.r.t. $\mathbf{D} \neq \mathbf{I}_n$.

4.2 Entropy Considerations on Localised Signals

We here add some considerations to the sparse and localised signal model of Section 2.1.2. Firstly, let the R.V. s have mean $\boldsymbol{\mu_s} = \mathbf{0}_n$ and covariance $\mathbf{K_s}$; consider $\mathbf{W}^S = \mathbf{A}^S \mathbf{D}$, yielding the R.V. of measurements $\mathbf{y}_S = \mathbf{W}^S \mathbf{s}$. This corresponds to a randomly chosen subset S of a full set of n measurements $\mathbf{y} = \mathbf{W}\mathbf{s}$, where the deterministic full-rank matrix $\mathbf{W} = \mathbf{AD}$. The measurements' covariance matrix $\mathbf{K}_{\mathbf{y}_S} = \mathbf{W}^S \mathbf{K_s} \left(\mathbf{W}^S\right)^*$ will be non-white in general, as localised s imply localised measurements. Moreover, $\forall S \subseteq \{0, \ldots, m-1\}$, $\mathbf{K}_{\mathbf{y}_S}$ are the principal submatrices of $\mathbf{K_y}$.

As an example, we may let $\mathbf{y}_S$ follow an anisotropic multivariate Gaussian distribution $\mathcal{N}(0, \mathbf{K}_{\mathbf{y}_S})$. Then localisation would directly indicate that the acquisition process represented by $\mathbf{W}^S$ is not maximising the quantity of information embedded in the measurements; this is seen by considering their *differential entropy*,

$$h(\mathbf{y}_S) = \int_{\mathbb{R}^m} f(\mathbf{y}_S) \log f(\mathbf{y}_S) \mathrm{d}\mathbf{y}_S \qquad (4.1)$$

as [130, Theorem 9.4.1] states that

$$h(\mathbf{y}_S) = \tfrac{1}{2} \log\left((2\pi e)^m \det \mathbf{K}_{\mathbf{y}_S}\right) \leq \tfrac{1}{2} \log\left(2\pi e \tfrac{\mathcal{E}_{\mathbf{y}_S}}{m}\right)^m \qquad (4.2)$$

which by [130, Theorem 16.8.4] attains the upper bound in (4.2) only when $\mathbf{y}_S$ is white for a fixed amount of average energy $\mathcal{E}_{\mathbf{y}_S} = \mathrm{tr}\left(\mathbf{K}_{\mathbf{y}_S}\right)$.

Clearly, the particular yet suggestive case of exactly Gaussian-distributed $\mathbf{y}_S$ would rigorously correspond to projecting $\mathbf{s} \sim \mathcal{N}(\mathbf{0}_n, \mathbf{K_s})$ over a deterministic $\mathbf{W}^S$; since this is not necessarily the case for k-sparse signals in conjunction with the use of PHE or PFE sensing matrices, the implicit yet reasonable hypothesis used here is that the distribution of $\mathbf{y}_S$ is well approximated by an elliptically-contoured one. Thus, the objective of our optimisation of the sensing matrix will substantially entail a selection of S that brings $h(\mathbf{y}_S)$ in the approximation of (4.2) as close as possible to its upper bound.

4.3 Maximum Entropy Sensing Matrix Designs for Compressed Sensing

The previous example indicates that the measurements $\mathbf{y}_S$ will not achieve the white-case entropy upper bound in (4.2) when the original signal is localised and $\mathbf{y}_S$ are obtained from sensing matrices $\mathbf{A}^S$ in the PHE (or similarly constructed ensembles). In such a constrained setting, we aim at finding the optimal $\mathbf{A}^S$ in the design space which conveys the maximum achievable quantity of information in the measurements $\mathbf{y}_S$.

4.3.1 A Maximum Entropy Problem

In general, we are searching for the m-cardinality subset $\mathbf{y}_{S^\star}$ of the full measurements' R.V. $\mathbf{y} = \mathbf{W}\mathbf{s}$ which attains the maximum differential entropy $h(\mathbf{y}_{S^\star})$, i.e., we solve

$$S^\star = \underset{S \subseteq \{0,\ldots,n-1\}, |S|=m}{\operatorname{argmax}} h(\mathbf{y}_S) \qquad (4.3)$$

In a Gaussian context, let $\mathbf{y} \sim \mathcal{N}(0, \mathbf{K_y})$ with $\mathbf{K_y} = \mathbf{W}\mathbf{K_s}\mathbf{W}^*$. Then $h(\mathbf{y}_S)$ is given in (4.2) where $\mathbf{K}_{\mathbf{y}_S}$ is a principal minor of $\mathbf{K_y}$ (i.e., $\mathbf{K}_{\mathbf{y}_S} = (\mathbf{K_y})_S^S = \mathbf{P}^S \mathbf{K_y} (\mathbf{P}^S)^*$) corresponding to the subset S of the full measurements' R.V. $\mathbf{y}$. Thus, if a Gaussianity hypothesis for $\mathbf{y}$ holds, (4.3) amounts to solving the following problem.

Problem 4.1 *(Maximum Entropy Sensing Matrix Design (deterministic rows case))*. Let $\mathbf{s} \in \mathbb{R}^n$ be a R.V. with covariance matrix $\mathbf{K_s}$; let $\mathbf{W} \in \mathbb{R}^{n \times n}$ be a deterministic full-rank matrix; let $\mathbf{y} = \mathbf{W}\mathbf{s}$ with covariance matrix $\mathbf{K_y}$. A design of $\mathbf{W}^S$ that maximises (4.1) is obtained by letting

$$S^\star = \underset{S \subseteq \{0,\ldots,n-1\}, |S|=m}{\operatorname{argmax}} \log \det \mathbf{K}_{\mathbf{y}_S} \qquad (4.4)$$

More realistically $\mathbf{y}$ will only be approximately Gaussian and substantially depends on the distribution of $\mathbf{s}$, thus requiring the more general solution of (4.3). While generally non-Gaussian, the density of s corresponding to natural signals decomposed on a sparsity basis is often approximated by multivariate Gaussian mixtures analogous to (2.1) for many natural signals such as images [145]. Differential entropy bounds tailored to the density of s would be required to rigorously expand (4.3) instead of (4.2); yet, bounding the differential entropy of such mixtures is an open problem [146].

In a distribution-agnostic fashion we still choose to solve problem (4.4): while the solution $S^\star$ might not achieve globally maximum entropy, by the well-known connection between Gaussian-case entropy maximisation and optimal linear prediction the corresponding $\mathbf{y}_{S^\star}$ will be the subset of measurements having least linear predictability (or equivalently maximum prediction error) from one another [147, Sec. 2.4.3 and 6.6]. Note that in Section 4.4 we also report reassuring evidence that natural signals considered in our experiments indeed produce approximately Gaussian $\mathbf{y}$, thus suggesting that (4.4) is well-posed for finding Hadamard sensing vectors that maximise the measurements' entropy.

4.3.2 A Heuristic Solution to Maximum Entropy Sensing Matrix Design

On the computational side, solving (4.4) amounts to finding the maximum-determinant principal minor $\mathbf{K}_{\mathbf{y}_{S^\star}}$. When $\mathbf{K}_\mathbf{y}$ is diagonal this is straightforwardly solved by

$$S^\star = \operatorname*{argmax}_{S \subseteq \{0,\ldots,n-1\}, |S|=m} \operatorname{tr}(\mathbf{K}_{\mathbf{y}_S})$$

i.e., by choosing the m largest-variance components of $\mathbf{y}$.

When this is not verified (4.4) is a well-known combinatorial problem. In [148] Ko *et al.* proved its hardness and proposed an exact branch-and-bound algorithm. Since in natural signals, fluctuations in $\mathbf{K_s}$ will occur and eventually require an update in $S^\star$ we propose to use a lightweight genetic algorithm [149] to find a heuristic solution to Problem 4.1.

The evolutionary analogue is obtained by mapping the i-th subset $S_{(i)}, |S_{(i)}| = m$ into a length n, binary-valued chromosome $\tau_{(i)} = I_{S_{(i)}}$

Procedure 1 Evolutionary Heuristic Solution of Problem 4.1.

Require: $\mathbf{K}_{n \times n}$ (covariance matrix), m, N_{gen}, N_{pop}, N_{par}, N_{chi}, P_{mut}
1: $\tau_{(0)} = I_{S_{(0)}} \in \Omega^{(0)}$, $S_{(0)} = \text{argmax}_{S \subseteq \{0,\ldots,n-1\}, |S|=m} \, \text{tr}\left(\mathbf{K}_S^S\right)$
2: **for all** $\tau_{(i)} \in \Omega^{(0)}, i > 0$ **do** {Random initialisation}
3:     Generate random $\tau_{(i)} = I_{S_{(i)}}$, $S_{(i)} \subseteq \{0, \ldots, n-1\}, |S_{(i)}| = m$.
4: **end for**
5: **for** $l = 0$ **to** $N_{\text{gen}} - 1$ **do** {Genetic search}
6:     **for all** $\tau_{(i)} \in \Omega^{(l)}$ **do** {Fitness evaluation}
7:         Calculate the fitness $f_{(i)} = 2\text{tr}\left(\log \text{diag}\left(\mathbf{L}_{S_{(i)}}\right)\right)$
8:     **end for**
9:     Sort $\tau_{(i)} \in \Omega^{(l)}$ in descending order w.r.t. their fitness $f_{(i)}$
10:     $\Omega_{\text{par}} \leftarrow \{\tau_{(i)}\}_{i=0}^{N_{\text{par}}-1}$ {Parents selection}
11:     **for** $k = 0$ **to** $N_{\text{chi}} - 1$ **do** {Mating phase}
12:         Randomly pick $\tau_{(i)}, \tau_{(j)} \in \Omega_{\text{par}}$
13:         $o \leftarrow$ random index in $\{1, \ldots, n-1\}$
14:         $\tau_{(N_{\text{par}}+k)} \leftarrow \left[(\tau_{(i)})_{0,\ldots,o-1} \, (\tau_{(j)})_{o,\ldots,n-1}\right]$ {Crossover}
15:         $r \leftarrow$ uniform random real in $[0, 1]$
16:         **if** $r < P_{\text{mut}}$ **then** {Mutation}
17:             Shuffle random $(0, 1)$ pairs in $\tau_{(N_{\text{par}}+k)}$
18:         **end if**
19:         **if** $|S_{(N_{\text{par}}+k)}| - m > 0$ **then** {Well-formed check}
20:             Remove $|S_{(i)}| - m$ exceeding ones in $(\tau_{(N_{\text{par}}+k)})_{0,\ldots,n-1}$
21:         **else**
22:             Add $m - |S_{(N_{\text{par}}+k)}|$ missing ones in $(\tau_{(N_{\text{par}}+k)})_{0,\ldots,n-1}$
23:         **end if**
24:     **end for**
25:     Eliminate duplicates and replenish the pool $\Omega^{(l+1)}$ {Uniqueness check}
26: **end for**
27: **return** $S^\star \leftarrow \text{supp} \, \tau_{(0)}, \tau_{(0)} \in \Omega^{(N_{\text{gen}}-1)}$

(the indicator function of $S_{(i)}$), whose fitness function is simply $f_{(i)} = \log \det \mathbf{K}_{\mathbf{y}_{S_{(i)}}}$. Since covariance matrices are symmetric PSD, by using the Cholesky factorisation $\log \det \mathbf{K}_{\mathbf{y}_{S_{(i)}}} = 2 \log \det \mathbf{L}_{S_{(i)}}$ where $\mathbf{L}_{S_{(i)}}$ is lower triangular. This allows fast and accurate computation of the i-th fitness $f_{(i)} = 2\text{tr}\left(\log \text{diag}\left(\mathbf{L}_{S_{(i)}}\right)\right)$.

The algorithm is implemented as in Proc. 1 for a generic covariance matrix $\mathbf{K}$ and controlled by the global parameters N_{gen} (number of generations), N_{pop} (population size at each generation). We note the use of a warm start by including in the initial population $\Omega^{(0)}$ the element $\tau_{(0)} = I_{S_{(0)}}$ initialised to the indices of the m largest variances in $\mathbf{K}$.

At each generation, mating occurs between the $N_{\text{par}} = \frac{N_{\text{pop}}}{3}$ highest-fitness parent chromosomes such that their $N_{\text{chi}} = \frac{2N_{\text{pop}}}{3}$

Procedure 2 Maximum Entropy Hadamard Sensing of Localised Signals.

1: Estimate $\widehat{\mathbf{K}}_\mathbf{y}$ and $\mathcal{E}_\mathbf{y} = \mathrm{tr}\left(\widehat{\mathbf{K}}_\mathbf{y}\right)$ (either by direct observation or by setting $\widehat{\mathbf{K}}_\mathbf{y} = \mathbf{W}\widehat{\mathbf{K}}_\mathbf{s}\mathbf{W}^*$ with $\widehat{\mathbf{K}}_\mathbf{s}$ the sparse coefficients' sample covariance matrix)
2: Evaluate $\overline{\mathbf{K}}_\mathbf{y} = (1-\gamma)\widehat{\mathbf{K}}_\mathbf{y} + \gamma\frac{\mathcal{E}_\mathbf{y}}{\sqrt{n}}\mathbf{I}_n$
3: Solve Problem 4.1 by running Proc. 1 on $\overline{\mathbf{K}}_\mathbf{y}$
4: Update the Hadamard sensing matrix $\mathbf{A}^{S^\star}$
5: **loop**
6:     Acquire $\mathbf{y}_{S^\star} = \mathbf{A}^{S^\star}\mathbf{x}$
7:     Recover $\hat{\mathbf{s}}$ by BP from $(\mathbf{A}^{S^\star}, \mathbf{y}_{S^\star})$
8: **end loop**
---

children are m-cardinality subsets. An elitist policy grants survival to the parent chromosomes until they are replaced by fitter children. To avoid possible stagnation in the population we have introduced common genetic operators such as one-point random crossover, random mutation with probability P_mut and a final uniqueness check to avoid clones in the population.

Thus, the algorithm yields a near-optimal $S^\star$ depending on the chosen parameters, which we use to construct a single $\mathbf{W}^{S^\star} = \mathbf{A}^{S^\star}\mathbf{D}$. Setting large $(N_\mathrm{gen}, N_\mathrm{pop})$ increases the complexity of this procedure but typically leads to a larger fitness gap between any random index subset S and the final $S^\star$. Moreover, rather than using a single $S^\star$ one may choose S from the high-entropy final population $\Omega^{(N_\mathrm{gen}-1)}$, which we refer to as the MaxDet pool for the following experimental evaluation.

4.4 Performance Evaluation

We here assess the near-optimality of the MaxDet pool against random PHE sensing matrices by observing the $\mathrm{ASNR}_{\hat{\mathbf{x}},\mathbf{x}}$ (as in (1.32)) attained by BP from the corresponding measurements $\mathbf{y}_S$. The signal classes tested here are natural images and ECGs taken from public-domain databases. To efficiently recover these signals by BP we used SPGL$_1$, while identical results were obtained by linear programming in GUROBI [93].

m	MaxDet pool PHE	Random PHE	RBE
1024 ($n/4$)	**36.57**	1.51	20.63
1365 ($\lfloor n/3 \rfloor$)	**39.63**	2.89	26.08

Table 4.1: Comparison of PHE and RBE sensing matrices on the USPS handwritten digits dataset: $\mathrm{ASNR}_{\hat{x},x}$ over 20 sample images; 25 MaxDet pool PHE, 25 random PHE and 50 RBE sensing matrices.

The experiments are carried out by following the proposed Hadamard sensing design strategy as summarised in Proc. 2 with heuristic parameters $N_{\mathrm{gen}} = 200, N_{\mathrm{pop}} = 50, P_{\mathrm{mut}} = 0.1$. Due to the importance of correctly inferring $\mathbf{K_y}$ we note the use of the *shrinkage covariance estimator*

$$\overline{\mathbf{K}}_\mathbf{y} = \gamma \widehat{\mathbf{K}}_\mathbf{y} + (1-\gamma)\frac{\mathrm{tr}\left(\widehat{\mathbf{K}}_\mathbf{y}\right)}{n}\mathbf{I}_n, \gamma \in [0,1)$$

which safely balances the sample covariance matrix $\widehat{\mathbf{K}}_\mathbf{y}$ with the same-energy white case. This covariance estimator leads (with suitable γ) to a full-rank $\overline{\mathbf{K}}_\mathbf{y}$ in the presence of additive measurement noise and from limited or linearly-dependent instances of $\mathbf{y}$.

Practical applications will also require an update in $S^\star$ to track statistically significant variations in the statistics of $\mathbf{s}$ (or $\mathbf{x}$). This update will be triggered whenever the recovered sparse coefficients are classified as outliers w.r.t. $\mathbf{K_s}$. However, we leave the analysis of online updates to future investigations.

4.4.1 Handwritten digits

The first experiment is carried out on image samples from the USPS handwritten digits database [150]. We estimate $\widehat{\mathbf{K}}_\mathbf{s}$ on a training set of 2000 images resized to 64×64 pixel ($n = 4096$) and with $\mathbf{D}$ the 2D DCT on which on-average $k = 467$ coefficients represent 95% of the original images' energy, thus complying with a compressible signal model as in Definition 1.4.

To show that the corresponding $\mathbf{y}$ can be considered Gaussian, we collect the full set of n Hadamard transform coefficients $\mathbf{y}$ for 4000

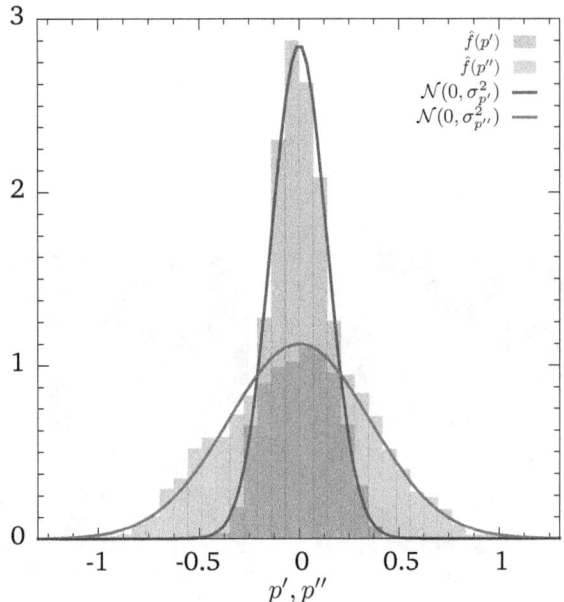

Figure 4.1: Empirical PDFs of p', p'' and their Gaussian approximation with $\sigma_{p'}^2 = 0.02, \sigma_{p''}^2 = 0.13$.

images in the database and project them on two randomly chosen, fixed orthonormal vectors $\mathbf{p}', \mathbf{p}'' \in \mathbb{R}^n$. The resulting empirical densities of $p' = (\mathbf{p}')^* \mathbf{y}, p'' = (\mathbf{p}'')^* \mathbf{y}$ are reported in Fig. 4.1 for 32 histogram bins and fit by univariate Gaussian distributions. Albeit with different variances, p', p'' are very well-fit by a Gaussian PDF; since these finite-n linear combinations of the coefficients in $\mathbf{y}$ are approximately Gaussian, as a result it is sensible that $\mathbf{y}$ has an approximately Gaussian behaviour and thus the entropy of $\mathbf{y}_S$ can be maximised by solving Problem 4.1.

By running Proc. 2 (1:-4:) with $\gamma = 10^{-12}$ we obtain MaxDet pools $\Omega^{(N_{\text{gen}}-1)}$ and near-optimal solutions S^* yielding high-entropy measurements for $m = 1024, 1365$. For each of 20 test images, we simulate CS by three sets of sensing matrices: 25 PHE matrices from the MaxDet pool (including the optimal $\mathbf{A}^{S^*}$), 25 randomly chosen PHE matrices and 50 RBE matrices. Signal recovery is then performed by BP from these sets of measurements. The results in terms of $\text{ASNR}_{\hat{\mathbf{x}}, \mathbf{x}}$

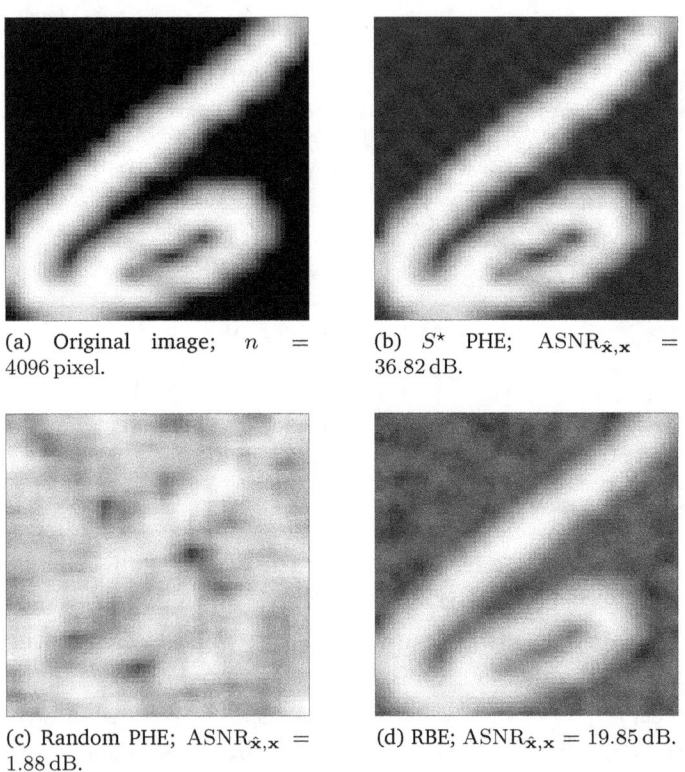

(a) Original image; $n = 4096$ pixel.

(b) $S^\star$ PHE; $\text{ASNR}_{\hat{\mathbf{x}},\mathbf{x}} = 36.82\,\text{dB}$.

(c) Random PHE; $\text{ASNR}_{\hat{\mathbf{x}},\mathbf{x}} = 1.88\,\text{dB}$.

(d) RBE; $\text{ASNR}_{\hat{\mathbf{x}},\mathbf{x}} = 19.85\,\text{dB}$.

Figure 4.2: Signal recovery of handwritten digits from maximum and non-maximum entropy sensing matrices; in the reported case $n = 4096, m = 1024$.

m	MaxDet pool PHE	Random PHE	RBE
64 ($n/4$)	**15.12**	2.68	6.94
85 ($\lfloor n/3 \rfloor$)	**17.20**	3.73	11.11

Table 4.2: Comparison of PHE and RBE sensing matrices on the PhysioNet ECG dataset: $\mathrm{ASNR}_{\hat{x},x}$ over 20 sample ECG; 25 MaxDet pool PHE, 25 random PHE and 50 RBE sensing matrices.

are documented in Table 4.1, where measurements obtained by the MaxDet pool Hadamard sensing matrices outperform both randomly selected PHE and RBE sensing matrices. Fig. 4.2 illustrates this observable improvement in terms of typical $\mathrm{ASNR}_{\hat{x},x}$ performances for a sample digit in the dataset and $m = 1024$.

4.4.2 Electrocardiographic tracks

In this second experiment we illustrate Hadamard sensing of ECG tracks from the PhysioNet database [120]. $\widehat{\mathbf{K}}_{\mathbf{s}}$ is estimated on a training set of 180 ECG fragments of dimensionality $n = 256$ and with $\mathbf{D}$ the Coiflet-3 orthonormal DWT [52], on which on-average $k = 39$ coefficients represent 95% of the original signal energy, thus verifying a compressible signal model. Although we omit their histogram, $\mathbf{y}$ can also be considered approximately Gaussian in this case.

Proc. 2 (1:-4:) with $\gamma = 10^{-12}$ yields MaxDet pools $\Omega^{(N_{\mathrm{gen}}-1)}$ and near-optimal solutions S^* for $m = 64, 85$ (a strongly undersampled setting). Signal recovery is then performed in the same fashion of the first example for 50 sample ECGs. The results in terms of $\mathrm{ASNR}_{\hat{x},x}$ are reported in Tab. 4.2, while Fig.4.3 illustrates the typical $\mathrm{ASNR}_{\hat{x},x}$ performances for a sample ECG in the dataset and $m = 85$.

From this further experimental evidence, we can conclude that the method described in this Chapter improves by a non-negligible amount the information conveyed by a localised signal into the measurements, allowing the selection of an optimal subset of the $\binom{n}{m}$ matrices that can be randomly constructed from the Hadamard matrix $\mathbf{H}_n$.

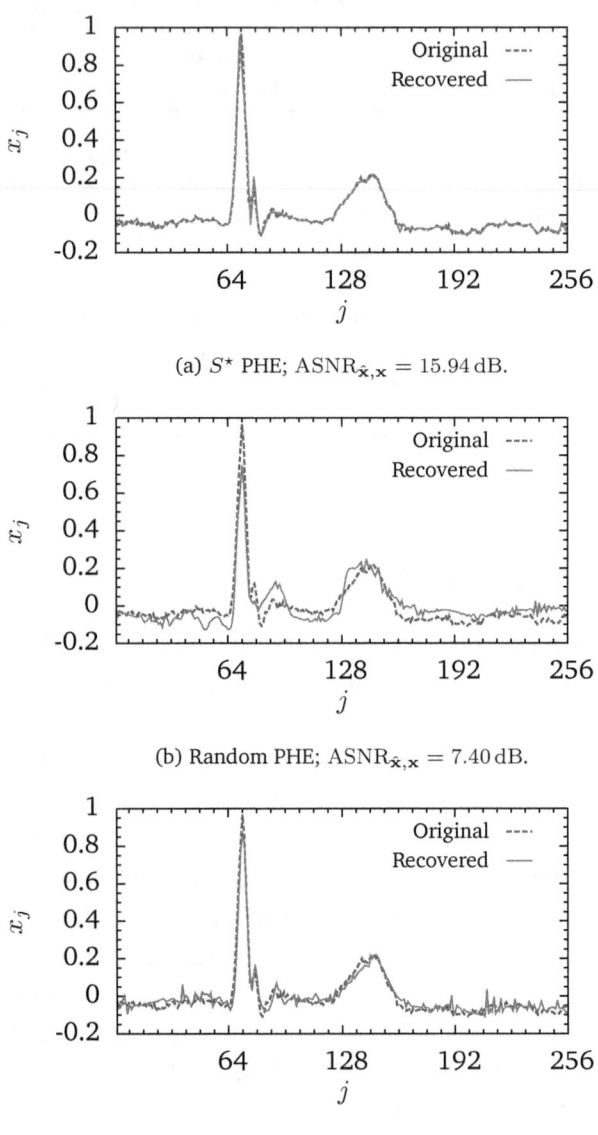

(a) $S^\star$ PHE; $\mathrm{ASNR}_{\hat{\mathbf{x}},\mathbf{x}} = 15.94\,\mathrm{dB}$.

(b) Random PHE; $\mathrm{ASNR}_{\hat{\mathbf{x}},\mathbf{x}} = 7.40\,\mathrm{dB}$.

(c) RBE, $\mathrm{ASNR}_{\hat{\mathbf{x}},\mathbf{x}} = 12.85\,\mathrm{dB}$

Figure 4.3: Signal recovery of ECGs from maximum and non-maximum entropy sensing matrices; in the reported case $n = 256, m = 85$.

Summary

- ▶ The design of RMEs from a finite set of deterministic sensing vectors (*i.e.*, a deterministic RME) matches some applications in which they are hard-coded in the analog or optical domain. This problem is of particular concern when the sparsity dictionary of the signal being acquired is correlated w.r.t. the sensing vectors.

- ▶ When the signals being acquired by CS are localised, an adaptive sensing matrix design criterion for this setting was proposed; the proxy for "information extraction" in this context is a maximisation of the measurements' differential entropy. In the context of choosing from a finite set of deterministic sensing vectors, this amounts to solving a maximum entropy problem by selecting an optimal set of row vectors to maximise this proxy.

- ▶ Due to the computational hardness of this maximisation, we have applied a heuristic algorithm to choose near-optimal sensing matrices in the corresponding deterministic ensemble. Experiments on two signal classes have shown that measurements provided by such matrices yield positive signal recovery performance increments w.r.t. randomly selected matrix designs.

II

Low-Complexity Security by Compressed Sensing

AVERAGE RECOVERY PERFORMANCES IN THE PRESENCE OF RANDOM MATRIX PERTURBATIONS

THE sensitivity of recovery algorithms w.r.t. a perfect knowledge of the sensing matrix is a general issue in many application scenarios in which CS is an option to acquire, or actually encode by means of low-complexity operations as in Chapter 3, natural signals complying with a sparse signal model.

Quantifying this sensitivity in order to predict the result of signal recovery is therefore valuable when no *a priori* information can be exploited, *e.g.*, when the encoding matrix is randomly perturbed without any exploitable structure. In this Chapter we focus on this aspect by means of a simplified least-squares model for the signal recovery problem, which enables the derivation of its average performance estimate that depends only on the interaction between the encoding and perturbation matrices.

The effectiveness and stability of the resulting heuristic in CS configurations where this evaluation is meaningful is demonstrated by numerical exploration of signal recovery under three simple random perturbation matrix models in a variety of cases; the aim of this treatment is to develop a sense of the fact that this observation can be leveraged to introduce a CS-based data protection scheme that exploits this sensitivity to missing information.

5.1 Compressed Sensing in the Presence of Perturbations

In some applications the knowledge of the sensing (or encoding) matrix may be imperfect due to perturbations; these may either be caused by the nature of the physical mechanism by which the sensing matrix is applied [151, 152] or to intentionally missing information at the decoder, *i.e.*, when the latter only knows the encoding matrix up to a certain degree of accuracy. In particular, while calibration [153, 154] may be attempted in the first case, the latter case could be designed to minimise the possibility of recovering missing information on the encoding matrix (as will, in fact, be done in Chapter 6). Thus, understanding the effects of a perturbation in the encoding matrix is a valuable information in most applications of CS.

5.1.1 Recovery Guarantees with Matrix Perturbations

We here assume that the encoding matrix can be decomposed as[1] $\mathbf{A}^{(1)} = \mathbf{A}^{(0)} + \mathbf{\Delta A}$, where $\mathbf{A}^{(0)} \in \mathbb{R}^{m \times n}$ is known to the decoder while $\mathbf{\Delta A} \in \mathbb{R}^{m \times n}$ is a perturbation matrix; in this case any clue on $\mathbf{\Delta A}$ is generally unavailable, and the corresponding term of $\mathbf{y} = \mathbf{A}^{(1)}\mathbf{x} = \mathbf{A}^{(0)}\mathbf{x} + \mathbf{\Delta A}\mathbf{x}$ is signal-dependent noise. We here let $\mathbf{D}$ be a known ONB that is available to the decoder, leaving the uncertainty to a perturbation matrix $\mathbf{\Delta A}$ and the actual sparse vector $\hat{\mathbf{s}}$ so that $\hat{\mathbf{x}} = \mathbf{D}\hat{\mathbf{s}}$.

In terms of evaluating the effect of such matrix perturbations a first fundamental result was given by Herman *et al.* [91], extending the established theoretical signal recovery guarantees for convex optimisation [13] to such perturbed cases; the following definition is required for a summary of this result.

---

[1]The reason for the notation $^{(1)}$ and $^{(0)}$ will be cleared out later; here it substantially indicates two levels of information, where the former denotes perfect knowledge of the truth and the latter a partial version of it.

Definition 5.1 *(Perturbation Constants [91])*. Let

$$\sigma^{(k)}_{\min/\max}(\mathbf{A}) = \min_{T\subseteq\{0,\ldots,n-1\},|T|=k}/\max \; \sigma_{\min/\max}(\mathbf{A}_T)$$

denote the extreme singular values among all k-column submatrices of a matrix $\mathbf{A} \in \mathbb{R}^{m\times n}$. We define the perturbation constants

$$\epsilon^{(k)}_{\mathbf{A}^{(1)}} \geq \frac{\sigma^{(k)}_{\max}(\mathbf{\Delta A D})}{\sigma^{(k)}_{\max}(\mathbf{A}^{(1)}\mathbf{D})}$$

$$\epsilon_{\mathbf{A}^{(1)}} \geq \frac{\sigma_{\max}(\mathbf{\Delta A D})}{\sigma_{\max}(\mathbf{A}^{(1)}\mathbf{D})} \geq \epsilon^{(k)}_{\mathbf{A}^{(1)}} \tag{5.1}$$

The modification of Theorem 1.9 is here rephrased for the recovery of k-sparse vectors in absence of other noise sources.

Theorem 5.1 *(Stable Recovery by BPDN in the Presence of Perturbations [91, Theorem 2])*. Let $\mathbf{y} = (\mathbf{A}^{(0)} + \mathbf{\Delta A})\mathbf{x} \in \mathbb{R}^m$ be noisy measurements with additive perturbation noise $\mathbf{\Delta A x} \in \mathbb{R}^m$; $\mathbf{x} = \mathbf{D s}$ with $\mathbf{D}$ an ONB and $\mathbf{s} \in \mathbb{R}^n : \|\mathbf{s}\|_0 = k$; $\mathbf{A}^{(1)} = \mathbf{A}^{(0)} + \mathbf{\Delta A} \in \mathbb{R}^{m\times n}$ verify the RIP with constant $\delta_{2k} < \sqrt{2}\left(1 + \epsilon^{(2k)}_{\mathbf{A}^{(1)}}\right)^{-2} - 1$ and $\epsilon^{(2k)}_{\mathbf{A}^{(1)}} < 2^{\frac{1}{4}} - 1$. Then $\hat{\mathbf{s}} = \Delta_{\mathrm{BPDN}}(\mathbf{y}, \mathbf{A}^{(0)}\mathbf{D}, \gamma)$ with noise threshold

$$\gamma = \epsilon^{(k)}_{\mathbf{A}^{(1)}} \sqrt{\frac{1+\delta_k}{1-\delta_k}} \|\mathbf{y}\|_2$$

is so that

$$\|\hat{\mathbf{s}} - \mathbf{s}\|_2 \leq c'_1 \gamma \tag{5.2}$$

where

$$c'_1 = \frac{4\sqrt{1+\delta_{2k}}\left(1+\epsilon^{(2k)}_{\mathbf{A}^{(1)}}\right)}{1-(\sqrt{2}+1)\left[(1+\delta_{2k})\left(1+\epsilon^{(2k)}_{\mathbf{A}^{(1)}}\right)^2 - 1\right]} \tag{5.3}$$

While formally correct, as in most other analyses based on the RIP once the error norm bound in (5.2) is actually computed the typical performances of signal recovery are much more accurate than those anticipated by Theorems 1.9, 5.1. In particular, the average MSE or

SNR performances rather than their worst-case upper bounds could be a useful tool in prospective applications (provided that the variance of such quantities is also limited). While prior works exist exploring the MSE lower bound for sparse signal recovery [155, 156] the particular case of matrix perturbations is covered in some contributions [157, 158] we here seek a design criterion following the principle that, in the generally complex behaviour of sparse signal recovery algorithms, an approach based on sensible mathematical intuition and sufficiently motivated by simulation delivers applicable results.

5.1.2 Signal Recovery Algorithms with Matrix Perturbations

With the basic information that noise is present on the measurements, a decoder may either:

1. choose to be *naive* and estimate $\hat{s} = \Delta_{\mathrm{BP}}(\mathbf{y}, \mathbf{A}^{(0)}\mathbf{D})$ in the erroneous assumption that the measurements are not affected by noise, forcing $\mathbf{y} = \mathbf{A}^{(0)}\hat{\mathbf{x}}$;

2. in a more informed fashion, attempt to guess a noise threshold ε so that $\hat{s} = \Delta_{\mathrm{BPDN}}(\mathbf{y}, \mathbf{A}^{(0)}\mathbf{D}, \varepsilon)$ actually attempts denoising with the prior information that the solution is sparse. The noise threshold must be set so that the norm $\|\Delta\mathbf{A}\mathbf{x}\|_2 \leq \varepsilon$. In a particularly optimistic case, the actual norm $\varepsilon^* = \|\Delta\mathbf{A}\mathbf{x}\|_2$ is here assumed to be known in the usual "genie"-tuning fashion;

3. in an ideal setting be provided with the actual support of s in $\mathbf{D}, \overline{T}$, so that it may estimate $\hat{s} = \Delta_{\mathrm{OLS}}(\mathbf{y}, \mathbf{A}^{(0)}\mathbf{D}, \overline{T})$. Note that this *non-perfectly informed* oracle solution is missing any prior on $\Delta\mathbf{A}$, therefore yielding the solution $\hat{s}$ that minimises the amount of error w.r.t. the components in $\overline{T}$.

A variety of algorithms and problem formulations tackle the general case of signal recovery under perturbations [159, 160], where significant improvements are therein shown to be possible when some structure in $\Delta\mathbf{A}$ can be leveraged. However, we explicitly focus on the case in which $\mathbf{\Delta A}$ is drawn from a RME with i.i.d. entries that changes at each instance of x; as noted in [160], signal recovery performances in this case are substantially limited to those of the aforementioned non-perfectly informed OLS.

5.2 An Average Performance Estimate in the Presence of Perturbations

The relative sophistication of Problems 1.5, 1.6 prevents an average analysis of the sensitivity w.r.t. the perturbation matrix in typical recovery problems. For this reason, in a simplified model we assume that (i) $(\mathbf{A}^{(0)}, \mathbf{\Delta A})$ are drawn from two RMEs with known and i.i.d. distributions of entries, (ii) an approximation of $\hat{\mathbf{x}} = \mathbf{D}\hat{\mathbf{s}}$ is obtained by solving $\Delta_{\mathrm{BP}}(\mathbf{y}, \mathbf{A}^{(0)}\mathbf{D})$ to satisfy $\mathbf{y} = \mathbf{A}^{(0)}\hat{\mathbf{x}}$. Pairing this with the original $\mathbf{y} = \mathbf{A}^{(1)}\mathbf{x}$ and with $\mathbf{\Delta A} = \mathbf{A}^{(1)} - \mathbf{A}^{(0)}$ we obtain

$$\mathbf{A}^{(0)}\mathbf{\Delta x} = \mathbf{\Delta A}\mathbf{x}, \quad \mathbf{\Delta x} = \hat{\mathbf{x}} - \mathbf{x} \qquad (5.4)$$

Starting from this, we further assume that $\mathbf{\Delta A}$ is indeed a *perturbation*, *i.e.*, that its entity is small w.r.t. $\mathbf{A}^{(0)}$. In this way, the least-squares approximation error $\mathbf{\Delta x}$ is supposed to be small, so we could assume that $\hat{\mathbf{x}}$ lies in a ball centred on x, and minimise its radius under the constraint (5.4), yielding the LLS solution

$$\mathbf{\Delta x} = \operatorname*{argmin}_{\mathbf{\Delta \zeta} \in \mathbb{R}^n} \|\mathbf{\Delta \zeta}\|_2^2 \text{ s.t. } \mathbf{A}^{(0)}\mathbf{\Delta \zeta} = \mathbf{\Delta A}\mathbf{x} \qquad (5.5)$$

that is $\mathbf{\Delta x} = (\mathbf{A}^{(0)})^{\dagger}\mathbf{\Delta A}\mathbf{x}$. To investigate the expectation of $\mathbf{\Delta x}$ when considered as a R.V., *i.e.*, the MSE of such a solution, we may then compute

$$\begin{aligned}\mathbb{E}\left[\|\mathbf{\Delta x}\|_2^2\right] &= \operatorname{tr}(\mathbf{C}_{\mathbf{\Delta x}}) \\ &= \operatorname{tr}\left(\mathbb{E}_{\mathbf{A}^{(0)}, \mathbf{\Delta A}, \mathbf{x}}\left[(\mathbf{A}^{(0)})^{\dagger}\mathbf{\Delta A}\mathbf{x}\mathbf{x}^*\mathbf{\Delta A}^*\left[(\mathbf{A}^{(0)})^{\dagger}\right]^*\right]\right) \\ &= \operatorname{tr}\left(\mathbb{E}_{\mathbf{A}^{(0)}, \mathbf{\Delta A}}\left[(\mathbf{A}^{(0)})^{\dagger}\mathbf{\Delta A}\mathbf{C}_{\mathbf{x}}\mathbf{\Delta A}^*\left[(\mathbf{A}^{(0)})^{\dagger}\right]^*\right]\right)\end{aligned} \qquad (5.6)$$

in the assumption that $\mathbf{A}^{(0)}$ and $\mathbf{\Delta A}$ are drawn from RMEs that are independent of $\mathbf{x}$, so the ratio

$$\frac{\mathbb{E}\left[\|\mathbf{\Delta x}\|_2^2\right]}{\mathbb{E}\left[\|\mathbf{x}\|_2^2\right]} = \operatorname{tr}\left(\mathbb{E}_{\mathbf{A}^{(0)},\mathbf{\Delta A}}\left[(\mathbf{A}^{(0)})^\dagger \mathbf{\Delta A} \frac{\mathbf{C_x}}{\mathcal{E}_\mathbf{x}} \mathbf{\Delta A}^\dagger \left[(\mathbf{A}^{(0)})^\dagger\right]^\dagger\right]\right) \quad (5.7)$$

where the energy-normalised correlation matrix $\mathbf{C_x}/\mathcal{E}_\mathbf{x}$ takes into account the second-order moments of the signal to acquire. Since $\mathbf{D}$ is assumed an ONB we may adopt a sparse signal model where each of $\binom{n}{k}$ supports of $\mathbf{s}$ has the same probability (see Section 2.1.1), and its k non-zero components are i.i.d. zero-mean r.v.s. With this, the correlation matrix $\mathbf{C_s}/\mathcal{E}_\mathbf{s} = \frac{1}{n}\mathbf{I}_n$ and $\mathbf{C_x} = \mathbf{D C_s D^*} = \frac{\mathcal{E}_\mathbf{s}}{n}\mathbf{I}_n$. In this particular case, a simplified evaluation of the *mean-signal-to-noise ratio* of the recovery[2], $\operatorname{MSNR}_{\hat{\mathbf{x}},\mathbf{x}} = \frac{\mathbb{E}[\|\mathbf{x}\|_2^2]}{\mathbb{E}[\|\mathbf{\Delta x}\|_2^2]}$ due to a perturbation of the encoding matrix is possible, yielding

$$\operatorname{MSNR}_{\hat{\mathbf{x}},\mathbf{x}} = n\left[\operatorname{tr}\left(\mathbb{E}_{\mathbf{A}^{(0)},\mathbf{\Delta A}}\left[(\mathbf{A}^{(0)})^\dagger \mathbf{\Delta A}\mathbf{\Delta A}^*\left[(\mathbf{A}^{(0)})^\dagger\right]^*\right]\right)\right]^{-1} \quad (5.8)$$

The expectation on $\mathbf{A}^{(0)}$ and $\mathbf{\Delta A}$ depends on the CS configuration we are considering and may be effectively computed in an empirical fashion by Monte Carlo simulations for any RME of interest. On the other hand, the more suggestive and equivalent

$$\operatorname{MSNR}_{\hat{\mathbf{x}},\mathbf{x}} = \left(\mathbb{E}_{\mathbf{A}^{(0)},\mathbf{\Delta A}}\left[\frac{1}{n}\sum_{j=0}^{n-1}\left[\sigma_j\left((\mathbf{A}^{(0)})^\dagger \mathbf{\Delta A}\right)\right]^2\right]\right)^{-1} \quad (5.9)$$

links the expected performance to the average of the singular values of $(\mathbf{A}^{(0)})^\dagger \mathbf{\Delta A}$, yet it is much less attractive in terms of computational requirements for a numerical exploration.

Note that such an estimate has a number of clear limitations:

1. since it focuses on non-denoising recovery (*i.e.*, the solution of $\Delta_{\operatorname{BP}}(\mathbf{y}, \mathbf{A}^{(0)}\mathbf{D})$) it underestimates the attained recovery quality when the disturbance due to the perturbation can be compensated by the relative abundance of information on the problem

---

[2]Note that $\operatorname{MSNR}_{\hat{\mathbf{x}},\mathbf{x}} \neq \operatorname{ASNR}_{\hat{\mathbf{x}},\mathbf{x}}$, since the expectation of a ratio is not in general the ratio of the numerator and denominator's expectations. Nevertheless, this is a completely sensible performance index.

due to (i) the availability of a large number of measurements in excess of the minimum required for recovery (therefore allowing efficient denoising) and (ii) knowing each instance's error norm $\varepsilon^* = \|\mathbf{\Delta Ax}\|_2$ with which $\Delta_{\mathrm{BPDN}}(\mathbf{y}, \mathbf{A}^{(0)}\mathbf{D}, \varepsilon^*)$ may be solved;

2. the estimate will lose its validity for small values of m that do not allow an effective recovery, *i.e.*, when even the *perfectly informed* $\Delta_{\mathrm{BP}}(\mathbf{y}, \mathbf{A}^{(1)}\mathbf{D})$ fails. In this case it is not sensible to assume that either (1.21) or (1.23) identify a good approximation of the true signal; the intrinsic reason is that $\|\mathbf{\Delta x}\|_2$ is not small (as the least-squares hypothesis in the neighbourhood of **x** will not hold[3]) and the estimate will not yield a relevant prediction of the recovery quality.

Thus (5.8) and the more general (5.7) are expected to be most effective when m is sufficiently large, so that the phase transition of $\Delta_{\mathrm{BP}}(\mathbf{y}, \mathbf{A}^{(1)}\mathbf{D})$ to the "1" region has occurred (see Section 1.2.4) but not much larger than the minimum m required to achieve it. Actually, this is how efficient CS configurations will be designed and why (5.8) will match the examples presented below.

5.3 Performance Evaluation with Random Matrix Perturbations

In this Section we choose different RMEs from which $\mathbf{\Delta A}$ is drawn, and introduce the *projection-to-perturbation ratio*,

$$\mathrm{PPR}_{\mathbf{A}^{(0)}, \mathbf{\Delta A}} = \frac{\mathbb{E}[\|\mathbf{A}^{(0)}\|_F^2]}{\mathbb{E}[\|\mathbf{\Delta A}\|_F^2]}$$

indicating the relative average energy of $\mathbf{A}^{(0)}$ w.r.t. $\mathbf{\Delta A}$ to control its impact.

5.3.1 Perturbation Models

In particular, we here discuss three perturbation models:

---
[3]A more formal, yet similar intuition drives some of the considerations in [156], albeit addressing a slightly different estimation problem.

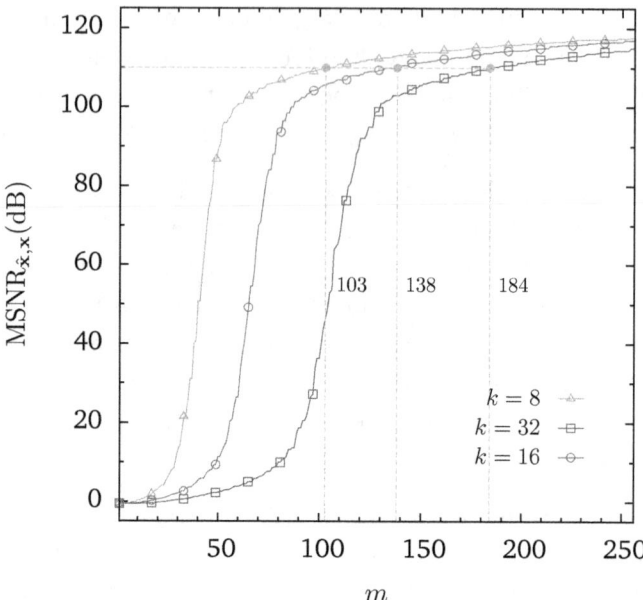

Figure 5.1: $\text{MSNR}_{\hat{x},x}$ curves used to set m beyond the phase transition of $\Delta_{\text{BP}}(y, A^{(1)}D)$.

1. *Dense Gaussian Additive (DGA)*: ΔA is drawn from the RGE with i.i.d. entries of variance $\sigma^2_{\Delta A} = \frac{1}{\text{PPR}_{A^{(0)},\Delta A}}$;

2. *Dense Uniform Multiplicative (DUM)*: $\Delta A = U \circ A^{(0)}$, where U is drawn from a RME that is independent of $A^{(0)}$ and has i.i.d. entries distributed as $U_{j,l} \sim \mathcal{U}\left(-\frac{\beta}{2}, \frac{\beta}{2}\right)$ and $\beta = 2\sqrt{\frac{3}{\text{PPR}_{A^{(0)},\Delta A}}}$;

3. *Sparse Sign-Flipping (SSF)*: a random set of index pairs C is independently generated so that each entry

$$\Delta A_{j,l} = \begin{cases} -2A^{(0)}_{j,l}, & (j,l) \in C \\ 0, & (j,l) \notin C \end{cases} \quad (5.10)$$

corresponds to a sign flipping of an element of $A^{(0)}$, where each pair of $\{0,\ldots,m-1\} \times \{0,\ldots,n-1\}$ has a probability η of being chosen. The resulting sparse RME has a density $\eta = \frac{1}{4\,\text{PPR}_{A^{(0)},\Delta A}}$ which controls $\sigma^2_{\Delta A} = 4\eta$.

5.3.2 Experiments and Estimates

In this numerical experiment we consider a simple setting of dimensionality $n = 256$ and assume $\mathbf{D}$ is the DCT; we generate s as a white R.V. by assuming equal probability of each of its $\binom{n}{k}$ possible supports, letting its k non-zero components be i.i.d. r.v.s distributed as $\mathcal{N}(0, 1/k)$. We consider $k = 8, 16, 32$ as prototypes of high-, medium-, and low-sparsity signals.

The matrix $\mathbf{A}^{(0)} \in \mathbb{R}^{m \times n}$ is here drawn from the RGE with unit-variance entries. As noted in the previous Section, we expect the estimate (5.8) to apply after a perfectly informed BP yields a solution with sufficiently large m.

For a quantitative evaluation of this aspect, we generate 200 instances of s, encode them with no perturbation and then apply $\Delta_{\mathrm{BP}}(\mathbf{y}, \mathbf{A}^{(1)}\mathbf{D})$ to measure the $\mathrm{MSNR}_{\hat{\mathbf{x}}, \mathbf{x}}$ with different values of m by means of SPGL_1. Given that the precision setting of the solver allows a maximum SNR of $\approx 120\,\mathrm{dB}$, by looking at the evidence in Fig. 5.1 we derive that a target $\mathrm{MSNR}_{\hat{\mathbf{x}}, \mathbf{x}}$ level of $110\,\mathrm{dB}$ is reached when $m = 103$ for $k = 8$, $m = 138$ for $k = 16$ and $m = 184$ for $k = 32$, at which it is safe to assume that the decoder is operating after the phase transition.

At these (m, k) pairs we explore the effect of perturbations and how closely it is predicted by (5.8); we choose the distribution parameters of the three models in Section 5.3.1 to obtain a given $\mathrm{PPR}_{\mathbf{A}^{(0)}, \boldsymbol{\Delta}\mathbf{A}} \in \{0, 5, \ldots, 80\}\,\mathrm{dB}$. For the chosen (m, k), we generate 200 instances of $(\mathbf{s}, \mathbf{A}^{(0)}, \boldsymbol{\Delta}\mathbf{A})$, encode $\mathbf{x} = \mathbf{D}\mathbf{s}$ with $\mathbf{A}^{(1)} = \mathbf{A}^{(0)} + \boldsymbol{\Delta}\mathbf{A}$ and attempt to recover $\hat{\mathbf{s}}$ by $\Delta_{\mathrm{BP}}(\mathbf{y}, \mathbf{A}^{(0)}\mathbf{D})$, $\Delta_{\mathrm{BPDN}}(\mathbf{y}, \mathbf{A}^{(0)}\mathbf{D}, \varepsilon^{\star})$ and the non-perfectly informed $\Delta_{\mathrm{OLS}}(\mathbf{y}, \mathbf{A}^{(0)}\mathbf{D}, \overline{T})$. These three results are compared with the outcome of a Monte Carlo simulation of our estimate in (5.8) averaged over 200 instances of $(\mathbf{A}^{(0)}, \boldsymbol{\Delta}\mathbf{A})$.

The results are depicted in Fig. 5.2b, 5.3b, 5.4b for fixed $k = 16$ and the three different perturbation models; the $\mathrm{MSNR}_{\hat{\mathbf{x}}, \mathbf{x}}$ of each decoder can be compared with the estimate as the $\mathrm{PPR}_{\mathbf{A}^{(0)}, \boldsymbol{\Delta}\mathbf{A}}$ increases (*i.e.*, the perturbation is made progressively smaller).

Moreover, since the estimate has negligible variations w.r.t. the perturbation model, we fix the latter to DGA and explore the effect of

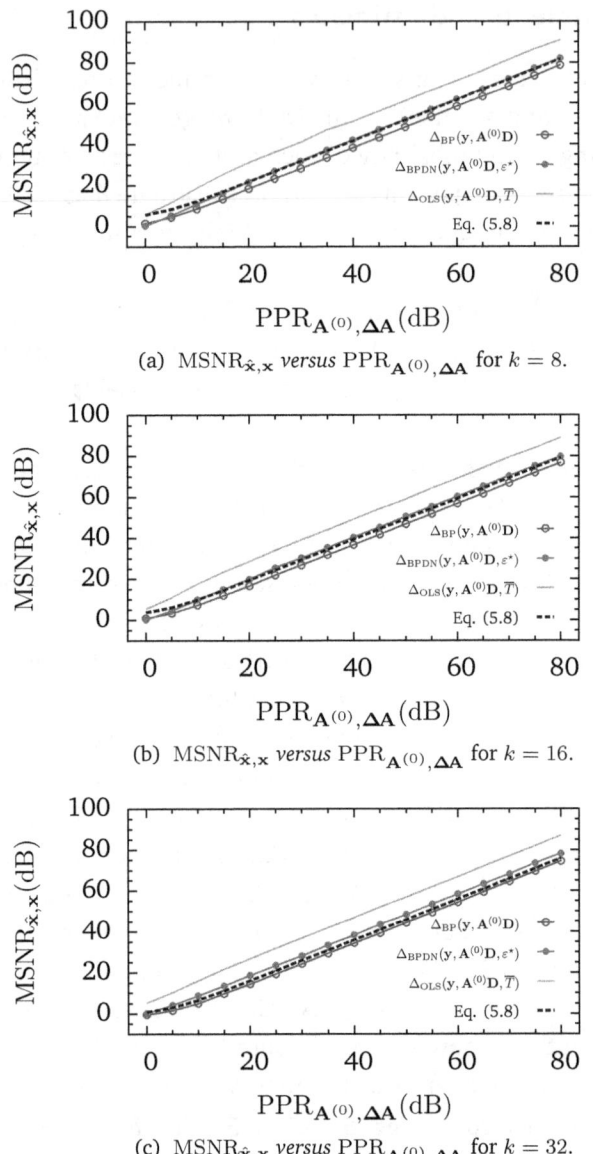

Figure 5.2: Comparison of the average performance estimate in (5.8) (dashed) against $\Delta_{\text{BP}}(\mathbf{y}, \mathbf{A}^{(0)}\mathbf{D})$ (empty circles), $\Delta_{\text{BPDN}}(\mathbf{y}, \mathbf{A}^{(0)}\mathbf{D}, \varepsilon^\star)$ (filled circles), $\Delta_{\text{OLS}}(\mathbf{y}, \mathbf{A}^{(0)}\mathbf{D}, \overline{T})$ (solid line) for the DGA perturbation.

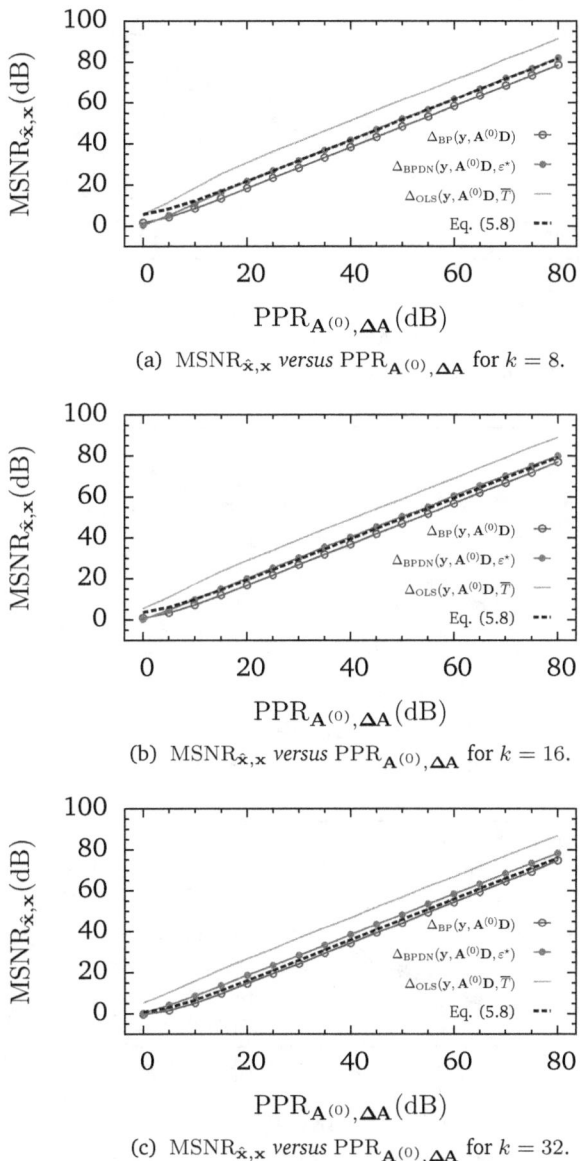

(a) $\mathrm{MSNR}_{\hat{\mathbf{x}},\mathbf{x}}$ versus $\mathrm{PPR}_{\mathbf{A}^{(0)},\Delta\mathbf{A}}$ for $k = 8$.

(b) $\mathrm{MSNR}_{\hat{\mathbf{x}},\mathbf{x}}$ versus $\mathrm{PPR}_{\mathbf{A}^{(0)},\Delta\mathbf{A}}$ for $k = 16$.

(c) $\mathrm{MSNR}_{\hat{\mathbf{x}},\mathbf{x}}$ versus $\mathrm{PPR}_{\mathbf{A}^{(0)},\Delta\mathbf{A}}$ for $k = 32$.

Figure 5.3: Comparison of the average performance estimate in (5.8) (dashed) against $\Delta_{\mathrm{BP}}(\mathbf{y},\mathbf{A}^{(0)}\mathbf{D})$ (empty circles), $\Delta_{\mathrm{BPDN}}(\mathbf{y},\mathbf{A}^{(0)}\mathbf{D},\varepsilon^\star)$ (filled circles), $\Delta_{\mathrm{OLS}}(\mathbf{y},\mathbf{A}^{(0)}\mathbf{D},\overline{T})$ (solid line) for the DUM perturbation.

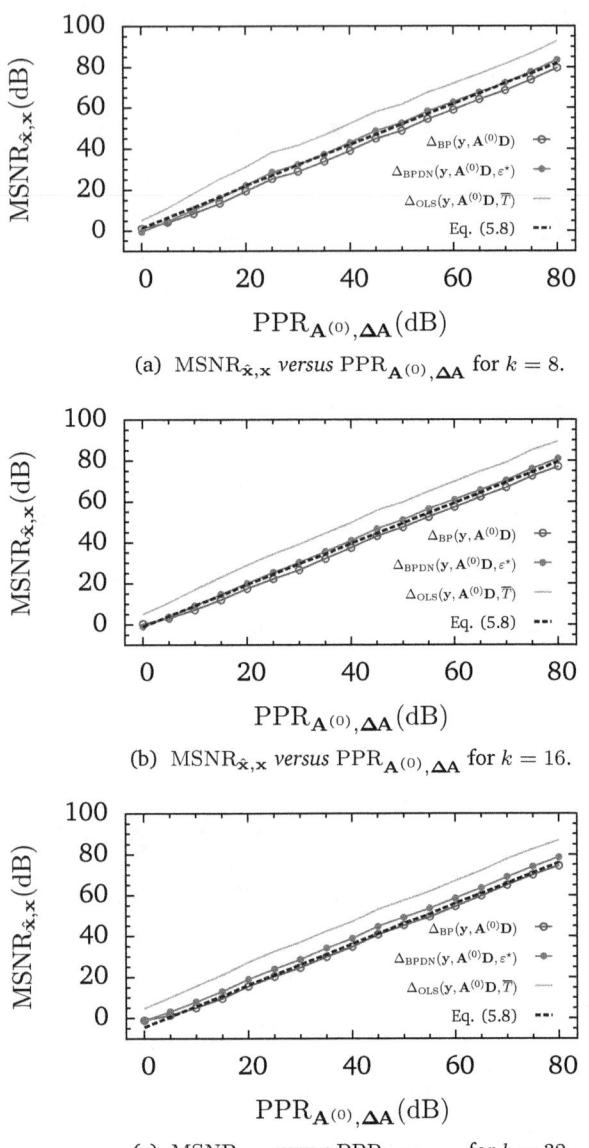

Figure 5.4: Comparison of the average performance estimate in (5.8) (dashed) against $\Delta_{\mathrm{BP}}(\mathbf{y}, \mathbf{A}^{(0)}\mathbf{D})$ (empty circles), $\Delta_{\mathrm{BPDN}}(\mathbf{y}, \mathbf{A}^{(0)}\mathbf{D}, \varepsilon^\star)$ (filled circles), $\Delta_{\mathrm{OLS}}(\mathbf{y}, \mathbf{A}^{(0)}\mathbf{D}, \overline{T})$ (solid line) for the SSF perturbation.

different sparsity levels at values for which the phase transition has occurred; the results are reported in Fig. 5.2a, 5.2b and 5.2c. Note that, although it is only an estimate, (5.8) appears to be quite effective in anticipating the average performances right between $\Delta_{\mathrm{BP}}(\mathbf{y}, \mathbf{A}^{(0)}\mathbf{D})$ and $\Delta_{\mathrm{BPDN}}(\mathbf{y}, \mathbf{A}^{(0)}\mathbf{D}, \varepsilon^*)$. This is coherent with its derivation that starts from a non-denoising, naive BP but assumes that the recovery has the ability of coming as close as possible to the true solution in the least-squares sense.

As a result of this performance estimate, we can conclude that the estimate in (5.8) (or its extension to non-white signals in (5.7)) is indeed sufficiently accurate to predict the average recovery performances of signal recovery algorithms under matrix perturbations right between $\Delta_{\mathrm{BP}}(\mathbf{y}, \mathbf{A}^{(0)}\mathbf{D})$ and $\Delta_{\mathrm{BPDN}}(\mathbf{y}, \mathbf{A}^{(0)}\mathbf{D}, \varepsilon^*)$, and in particular when the configuration of CS being used is operating in the appropriate region of the phase space, *i.e.*, when the set of (m, n, k) is so that recovery by BP is always feasible.

In the next Chapter, we will focus on SSF as a method to introduce a controlled, conveniently generated perturbation in an encoding matrix to deliver data protection embedded in the sensing or encoding process; hence the interest in the devised estimate, that serves as a design formula for a lightweight encryption protocol.

Summary

- Matrix perturbations have a strong effect on the recovery performances of BPDN, and quantifying their effect is important when they cannot be overcome by calibration procedures.

- An estimate of the average recovery performances attained by CS in the presence of a random matrix perturbation was proposed. Our performance index is simply calculated by estimating the effect of the perturbation on the singular values with Monte Carlo trials, and requires no prior information on the signal support. However, the performance estimate applies only after the phase transition of the corresponding recovery problem, *i.e.*, where CS will actually be used in practice.

▶ The estimate was shown to be numerically stable w.r.t. a variety of different matrix perturbations. One particular perturbation, *i.e.*, Sparse Sign-Flipping has a fundamental role in this thesis as a method to provide a multiclass encryption mechanism that will be devised in the following Chapter.

6

LOW-COMPLEXITY MULTICLASS ENCRYPTION BY COMPRESSED SENSING

With the rise of paradigms such as wireless sensor networks, whose lightweight node requirements have been anticipated in Chapter 3, the matter of defending the privacy of digital data gathered and distributed by such networks is a relevant issue, of even more concern when such networks acquire sensitive information such as biometric signals used in health monitoring applications. This data protection is normally granted by means of encryption stages securing the transmission channel [161] that is implemented in the digital domain after A/D conversion of the signal. Due to their complexity, cryptographic modules (*e.g.*, those implementing the *Advanced Encryption Standard (AES)* [162]) may require a considerable amount of resources, especially in terms of power consumption.

In this Chapter, we investigate the possibility of using CS with encoding matrices drawn from the RBE as a physical-layer method to embed security properties directly into the acquisition process. Such an almost-zero cost encryption mechanism is an appealing option that could be applied seamlessly in a digital-to-digital fashion, as it is perfectly integrated into the scheme of Fig. 3.2c.

Although it is well known that, due to its linearity, CS cannot be regarded as a mean to provide *perfect secrecy* [26], Chapters 7 and 8 will thoroughly address the fact that both computational and weak theoretical secrecy conditions hold.

The proposed encryption strategy relies on the fact that any receiver attempting to decode the reduced-dimensionality measurements produced by applying a RBE encoding matrix must know the true instance of the latter as used in the acquisition process to attain exact signal recovery. In absence of this information, it will suffer the equivalent of a matrix perturbation, whose decoder-side sensitivity has been anticipated by the analysis in Chapter 5. We will actually exploit this sensitivity to provide multiple recovery quality-based access levels, *i.e.*, *classes* of access to the information content embedded in the signal. In fact, when the *true encoding matrix* is completely unknown the signal is fully encrypted, whereas a receiver that knows the encoding matrix up to some random perturbations indeed will see the quality degradation anticipated in the previous Chapter.

We therefore aim to control the recovery performances of users (receivers) belonging to the same class by exploiting their ignorance of the true encoding matrix. Since these encoding matrices and the corresponding perturbations are generated from the available *private keys* at the corresponding decoders, *high-class* receivers are given a *complete key* and thus the true encoding matrix, while *lower-class* receivers are given an *incomplete key* resulting in a partially corrupted encoding matrix, with no clue on where the perturbations have occurred. To ensure that this mismatch is undetected by lower-class receivers, we will only alter the sign of a randomly chosen subset of the entries of the true encoding matrix, *i.e.*, we apply a SSF perturbation (as defined in the previous Chapter) that ensures undetectability since the encoding matrix is drawn from the RBE.

To quantify in a more precise manner the capabilities of the proposed multiclass encryption scheme we will (i) provide upper- and lower-bound analyses on the recovery error norm suffered by lower-class receivers depending on the chosen amount of perturbation (ii) evaluate exemplary applications of multiclass CS to concealing sensitive information in images, ECGs and speech signals. The attained recovery performances are also evaluated in a signal processing perspective by means of automatic feature-extraction algorithms to prove the efficacy of this strategy at integrating some security properties in the sensing or encoding process.

6.1 Principles of Multiclass Encryption by Compressed Sensing

We here discuss the introduction of a novel lightweight scheme for data protection, namely, *multiclass encryption* by CS.

6.1.1 Secrets, Bits and Compressed Sensing

Standard CS has been previously interpreted [26, 163] as a private-key cryptosystem where x is the *plaintext*, the measurement vector y is the *ciphertext* and the *encryption algorithm* is the linear transformation operated by the encoding matrix **A**. In the classic setting, Alice would acquire a plaintext x by CS using **A** and send to Bob the ciphertext y; Bob is therefore able to successfully recover x from y if he is provided with **A**, or equivalently the *encryption key* or *shared secret* required to generate it. In fact, CS may be regarded as similar to a Hill cipher [164, Section 2.7], where the main difference is the fact that the encoding matrix is non-invertible and decryption is actually conditional to (i) the sparsity of the plaintext w.r.t. a suitable basis (here assumed an ONB for the sake of simplicity) and (ii) the knowledge of **A**, that is necessary in the recovery of x from y.

Focusing on **A**, we know that any error in its entries reflects on the quality of the reconstructed signal as discussed in Chapter 5. Since private-key communications operate by agreeing on a finite number of bits that form the above secret, the mn symbols that comprise $\mathbf{A} \in \mathbb{R}^{m \times n}$ will be extracted from a pseudo-random sequence of bits that is obtained from the initial seed. In our case, the seed is given by the secret itself, and depending on the number of its bits the expanded pseudo-random binary sequence will eventually repeat. In the following we will assume that the period of the pseudo-random sequences generated by algorithmic expansion of the secret (*e.g.*, by a *Pseudo-Random Number Generator (PRNG)*) is sufficiently long as to

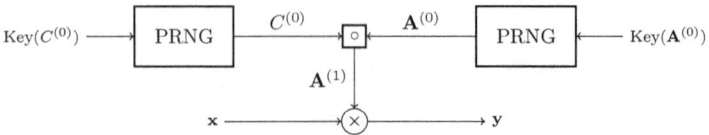

(a) The encoder; × here denotes the matrix-vector product, ○ the composition by (6.1).

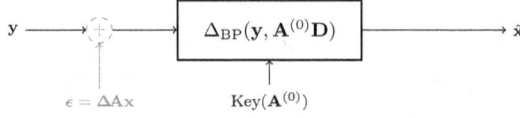

(b) A second-class decoder; the virtual effect of missing information on the encoding matrix at the decoder is highlighted in red.

Figure 6.1: An overview of two-class encryption by CS.

guarantee that in an observation time of interest the same $\mathbf{A}$ will never occur twice. This hypothesis is fundamental to ensure that $\mathbf{A}$ cannot be recovered from the knowledge of a sufficient number of plaintexts and ciphertexts, a simple observation first made by Drori [165] together with an analogy with the *one-time pad* [164, Section 2.9].

Since both the RGE and RBE are suitable ensembles to extract the encoding matrices we assume that the $\mathbf{A}$ of interest are all drawn from the RBE. This choice is motivated by the fact that the number of bits required to produce a single encoding matrix symbol is therefore maximised, as the bits output by the PRNG before repetition are a precious resource. With this hypothesis, we will let any instance of any RBE encoding matrix be a generic, unique element in a long-period repeatable sequence.

6.1.2 Multiclass Compressed Sensing

Let us now consider a scenario where multiple users receive the same ciphertext $\mathbf{y} = \mathbf{A}^{(1)}\mathbf{x}$, know the ONB $\mathbf{D}$ in which the plaintext $\mathbf{x}$ is k-sparse, but are made different by the fact that some of them know the true $\mathbf{A}^{(1)}$ whereas the others only know an approximate version of it, *i.e.*, $\mathbf{A}^{(0)}$. The resulting mismatch between $\mathbf{A}^{(1)}$ and $\mathbf{A}^{(0)}$, which will be used in the decoding process by the latter set of receivers, will limit the quality of their signal recovery as detailed below.

A Two-Class Encryption Scheme

With this principle in mind a straightforward, undetectable method to introduce controlled perturbations is flipping the sign of a subset of the entries of the encoding matrix in a random pattern. More formally, let $\mathbf{A}^{(0)} \in \{-1, +1\}^{m \times n}$ denote the *initial encoding matrix* and $C^{(0)}$ be a subset of $c < mn$ index pairs chosen at random for each $\mathbf{A}^{(0)}$. We therefore construct the *true encoding matrix* $\mathbf{A}^{(1)}$ by

$$\forall (j,l) \in \{0, \ldots, m-1\} \times \{0, \ldots, n-1\},$$

$$A_{j,l}^{(1)} = \begin{cases} A_{j,l}^{(0)}, & (j,l) \notin C^{(0)} \\ -A_{j,l}^{(0)}, & (j,l) \in C^{(0)} \end{cases} \quad (6.1)$$

and use it to encode $\mathbf{x}$ by $\mathbf{y} = \mathbf{A}^{(1)}\mathbf{x}$. Although this alteration simply involves inverting c randomly chosen sign bits in a buffer of mn pseudo-random symbols, we will use its linear perturbation model

$$\mathbf{A}^{(1)} = \mathbf{A}^{(0)} + \mathbf{\Delta A} \quad (6.2)$$

as in Chapter 5, where $\mathbf{\Delta A}$ is a c-sparse random matrix[1]

$$\forall (j,l) \in \{0, \ldots, m-1\} \times \{0, \ldots, n-1\},$$

$$\Delta A_{j,l} = \begin{cases} 0, & (j,l) \notin C^{(0)} \\ -2A_{j,l}^{(0)}, & (j,l) \in C^{(0)} \end{cases} \quad (6.3)$$

or equivalently

$$\forall (j,l) \in \{0, \ldots, m-1\} \times \{0, \ldots, n-1\},$$

$$\Delta A_{j,l} = \begin{cases} 0, & (j,l) \notin C^{(0)} \\ 2A_{j,l}^{(1)}, & (j,l) \in C^{(0)} \end{cases} \quad (6.4)$$

---

[1] To be specific, it can be seen as drawn from a ternary-valued RME $\mathbf{\Delta A} \in \{-2, 0, 2\}^{m \times n}$ constructed from all the equiprobable assignments of c non-zero elements verifying (6.3). In a simplifying view, we let it have i.i.d. entries $\forall (j,l) \in \{0, \ldots, m-1\}, \{0, \ldots, n-1\}, \mathbb{P}[\Delta A_{j,l} = -2] = \mathbb{P}[\Delta A_{j,l} = 2] = \frac{\eta}{2}, \mathbb{P}[\Delta A_{j,l} = 0] = 1 - \eta$, so the density parameter is actually controls the probability assignment.

with SSF density $\eta = \frac{c}{mn}$. By doing so, any receiver is still provided an encoding matrix differing from the true one by an instance of $\mathbf{\Delta A}$. This perturbation is *undetectable, i.e.*, $\mathbf{A}^{(1)}$ and $\mathbf{A}^{(0)}$ are statistically indistinguishable since they are equal-probability realisations of the same RBE, with all points in $\{-1,+1\}^{m \times n}$ having the same probability.

A *first-class* user receiving $\mathbf{y} = \mathbf{A}^{(1)}\mathbf{x} = (\mathbf{A}^{(0)} + \mathbf{\Delta A})\mathbf{x}$ and knowing $\mathbf{A}^{(1)}$ is therefore able to recover, in absence of other noise sources and with m sufficiently larger than the sparsity $k : \mathbf{x} = \mathbf{D}\mathbf{s}, \|\mathbf{s}\|_0 = k$, the exact sparse solution $\hat{\mathbf{s}} = \mathbf{s}$ by solving $\Delta_{\text{BP}}(\mathbf{y}, \mathbf{A}^{(1)}\mathbf{D})$. A *second-class* user only knowing $\mathbf{y}$ and $\mathbf{A}^{(0)}$ is instead subject to an equivalent signal- and perturbation-dependent, non-white noise term ϵ due to missing pieces of information on $\mathbf{A}^{(1)}$, that is

$$\mathbf{y} = \mathbf{A}^{(1)}\mathbf{x} = \mathbf{A}^{(0)}\mathbf{x} + \epsilon \qquad (6.5)$$

where $\epsilon = \mathbf{\Delta A}\mathbf{x}$ is a pure disturbance since both $\mathbf{\Delta A}$ and $\mathbf{x}$ are unknown to the second-class receiver. Its approximation $\hat{\mathbf{x}}$ is obtained the solution of, *e.g.*, $\Delta_{\text{BP}}(\mathbf{y}, \mathbf{A}^{(0)}\mathbf{D})$ or $\Delta_{\text{BPDN}}(\mathbf{y}, \mathbf{A}^{(0)}\mathbf{D}, \varepsilon^*)$ with $\varepsilon^* = \|\epsilon\|_2$, where the considerations made in Chapter 5 seamlessly apply; performing signal recovery in the erroneous assumption that $\mathbf{y} = \mathbf{A}^{(0)}\hat{\mathbf{x}}$, *i.e.*, with a corrupted encoding matrix will lead to a noisy $\hat{\mathbf{x}} = \mathbf{D}\hat{\mathbf{s}}$.

In terms of recovery guarantees, while upper bounds on the recovery error norm $\|\hat{\mathbf{x}} - \mathbf{x}\|_2$ have been anticipated in the form of Theorem 1.9 and 5.1, the crucial matter in this Chapter will be finding a lower bound to the error norm, *i.e.*, a *best-case analysis* of the second-class recovery error. We anticipate that this will depend on the perturbation density η, which will be suitably chosen to fix the desired quality range for each class. This is precisely obtained in Section 6.2, together with a quantification of the upper bound by a direct application of Theorem 5.1.

A Multiclass Encryption Scheme

The two-class scheme may be iterated to devise an arbitrary number of user classes: a SSF can be applied on disjoint subsets of index pairs

$C^{(u)}$, $u \in \{0, \ldots, w-2\}$ of $\mathbf{A}^{(0)}$ so that

$$A_{j,l}^{(u+1)} = \begin{cases} A_{j,l}^{(u)}, & (j,l) \notin C^{(u)} \\ -A_{j,l}^{(u)}, & (j,l) \in C^{(u)} \end{cases}$$

yielding the corresponding $\{\mathbf{A}^{(u)}\}_{u=0}^{w-1}$, each in turn associated with one of w user classes that progressively complete the knowledge of the true encoding $\mathbf{A}^{(w-1)}$. Thus, if the plaintext **x** is encoded with $\mathbf{A}^{(w-1)}$ we may distinguish *high-class* users knowing the complete encoding $\mathbf{A}^{(w-1)}$, *low-class* users knowing only $\mathbf{A}^{(0)}$ and *mid-class* users knowing $\mathbf{A}^{(u+1)}$ with $u = 0, \ldots, w-3$. This simple technique can be applied to provide multiple classes of access to the information in **x** granting different signal recovery performances at the decoder.

A System Perspective

The strategy described in this Section provides a multiclass encryption architecture where the shared secret between the CS encoder and each receiver is distributed depending on the quality level granted to the latter. In particular, the full encryption key of a w-class CS scheme is composed of w seeds, *i.e.*, low-class users are provided the secret $\text{Key}(\mathbf{A}^{(0)})$, class-1 users are provided $\text{Key}(\mathbf{A}^{(1)}) = (\text{Key}(C^{(0)}), \text{Key}(\mathbf{A}^{(0)}))$ up to high-class users with

$$\text{Key}(\mathbf{A}^{(w-1)}) = \left(\text{Key}(C^{(w-2)}), \cdots, \text{Key}(C^{(0)}), \text{Key}(\mathbf{A}^{(0)})\right)$$

An exemplary network implementing this policy is depicted in Fig. 6.3. This is reduced to the simple scheme of Fig. 6.1 in the case of a two-class encryption, where $\text{Key}(C^{(0)}), \text{Key}(\mathbf{A}^{(0)})$ fully define the key-agreement.

From the resources point of view, multiclass CS can be implemented with practically zero computational overhead. The encoding matrix generator is substantially a PRNG (*e.g.*, a *Linear Feedback Shift Register (LFSR)*) and is structurally identical at both the encoder and high-class decoder side, whereas lower-class decoders may use the same encoding matrix generation scheme but are unable to rebuild the true one due to the missing pieces of the shared secret, *i.e.*, $\text{Key}(C^{(u)})$.

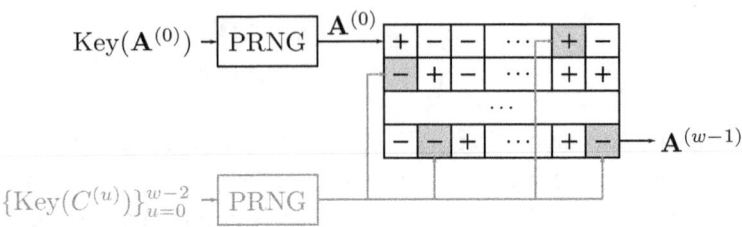

Figure 6.2: Encoding matrix generator architecture.

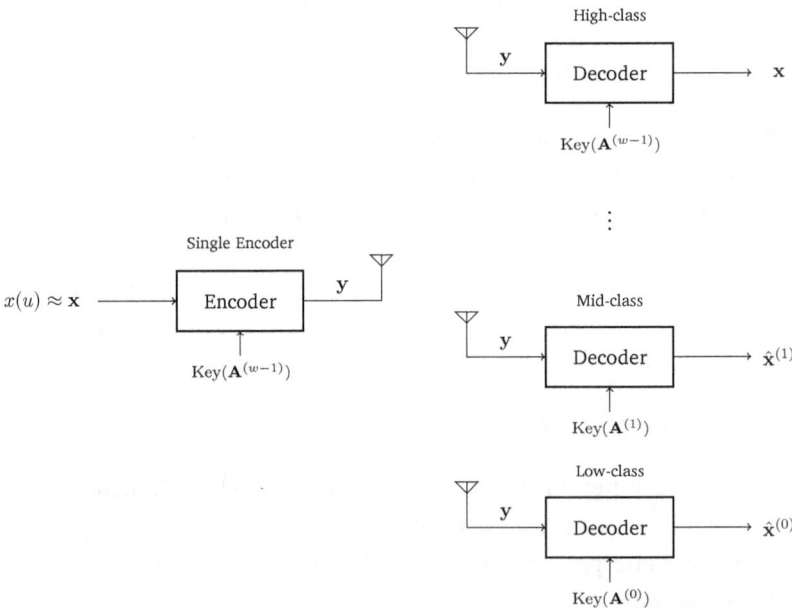

Figure 6.3: A single-transmitter, multiple-receiver multiclass CS network: the encoder acquires an analog signal $x(u)$ by CS and transmits the measurement vector $\mathbf{y}$. Low-quality decoders reconstruct a signal approximation with partial knowledge of the encoding, resulting in perturbation noise and leading to an approximate solution $\hat{\mathbf{x}}^{(u)}$ for the u-th user class.

The initial matrix $\mathbf{A}^{(0)}$ is, as anticipated, updated from a pseudo-random binary stream generated by expanding $\mathrm{Key}(\mathbf{A}^{(0)})$ with a PRNG. The introduction of sign-flipping is a simple post-processing step carried out on the stream buffer by reusing the same PRNG architecture and expanding the corresponding $\mathrm{Key}(C^{(u)})$, thus having minimal computational cost (see Fig. 6.2). Of course, the PRNGs have to be carefully chosen to avoid cryptanalysis [166]; however, since the values generated by this PRNG are never exposed, cryptographically-secure PRNGs [167] or security-enhancing primitives on the output [168] may be avoided to save resources, provided that the period with which the matrices are reused is kept sufficiently large.

6.2 Recovery Error Guarantees and Bounds

We now analyse the properties of multiclass CS starting from some statistical priors on the signal being encoded. Rather than relying on its *a priori* distribution, our analysis uses general moment assumptions that may correspond to many probability distributions on the signal domain. We will therefore adopt the following models:

(m_1) for finite n, we let $\mathbf{x} = \{x_j\}_{j=0}^{n-1}$ be a real-valued R.V.. Its realisations are *finite-length* plaintexts denoted by the same letter $\mathbf{x}$, and are assumed to have finite energy $E_\mathbf{x} = \|\mathbf{x}\|_2^2$. We will let each $\mathbf{x} = \mathbf{D}\mathbf{s}$ with $\mathbf{D}$ an ONB and $\mathbf{s}$ being k-sparse to comply with sparse signal recovery guarantees. $\mathbf{x}$ is mapped to the measurements' R.V. $\mathbf{y} = \{y_j\}_{j=0}^{m-1}$ whose realisations are the finite-length ciphertexts denoted by the same letter $\mathbf{y}$ as $\mathbf{y} = \mathbf{A}^{(1)}\mathbf{x}$;

(m_2) for $n \to \infty$, we let $\mathcal{X} = \{x_j\}_{j=0}^{+\infty}$ be a real-valued *Random Process* (R.P.). Its realisations or *infinite-length* plaintexts $\mathbf{x}$ are assumed to have finite power $W_\mathbf{x} = \lim_{n\to\infty} \frac{1}{n}\sum_{j=0}^{n-1} x_j^2$. We may denote them as sequences $\mathbf{x} = \{\mathbf{x}^{(n)}\}_{n=0}^{+\infty}$ of finite-length plaintexts $\mathbf{x}^{(n)} = \begin{bmatrix} x_0 & \cdots & x_{n-1} \end{bmatrix}$. $\mathcal{X}$ is mapped to either a R.V. $\mathbf{y}$ of finite-length ciphertexts for finite m, or a R.P. $\mathcal{Y} = \{y_j\}_{j=0}^{+\infty}$ of infinite-length ciphertexts for $m, n \to \infty$, $\frac{m}{n} \to q$. Both cases are comprised of[2] r.v.s $y_j = \frac{1}{\sqrt{n}} \sum_{l=0}^{n-1} A_{j,l}^{(1)} x_l$.

When none of the above models is specified, a single instance of $\mathbf{y} = \mathbf{A}^{(1)}\mathbf{x}$ is considered as in the standard CS framework.

In order to quantify the recovery quality performance gap between low- and high-class users receiving the same measurements $\mathbf{y} = \mathbf{A}^{(1)}\mathbf{x}$, we now provide performance bounds on the recovery error in the simple two-class case, starting from the basic intuition that if the sparsity basis of $\mathbf{x}$ is not the canonical basis, then most plaintexts $\mathbf{x} \notin \mathrm{Ker}(\Delta \mathbf{A})$ so the perturbation noise $\boldsymbol{\epsilon} \neq \mathbf{0}_m$.

6.2.1 Second-Class Recovery Error Norm: Lower Bound

The following results aim at predicting the best-case recovery quality of any second-class decoder that assumes $\mathbf{y}$ was encoded by $\mathbf{A}^{(0)}$, whereas $\mathbf{y} = \mathbf{A}^{(1)}\mathbf{x}$ in absence of other noise sources and regardless of the sparsity of $\mathbf{x}$. Since $\mathbf{A}^{(1)}$ is drawn from the RBE any exact signal recovery guarantee based on the properties of this matrix ensemble holds when encoding $\mathbf{x}$ by $\mathbf{A}^{(1)}$. By such guarantees, the dimensionality m of the measurement vector $\mathbf{y}$ must exceed the sparsity k by a quantity depending on the rate k/n. In the following, we will assume that m/n and k/n grant that a decoder knowing the true encoding $\mathbf{A}^{(1)}$ is able to accurately reconstruct the original signal by $\Delta_{\mathrm{BP}}(\mathbf{y}, \mathbf{A}^{(1)}\mathbf{D})$.

We now introduce a result that shows how the recovery error norm suffered by a second-class receiver is *at least* (rather than *at most*, as is usually the case for performance guarantees in CS) a certain quantity essentially depending on the nature of the perturbation $\Delta \mathbf{A} = \mathbf{A}^{(1)} - \mathbf{A}^{(0)}$; this will serve as a basic design guideline for multiclass encryption schemes.

---

[2]The $1/\sqrt{n}$ scaling is not only theoretically needed for normalisation purposes, but also practically required in the design of quantiser ranges.

Recovery Error Guarantees and Bounds

Theorem 6.1 *(Second-class recovery error lower bound (non-asymptotic case))*. Let:

- $\mathbf{A}^{(0)}, \mathbf{A}^{(1)} \in \{-1, +1\}^{m \times n}$ be drawn from the RBE and $\mathbf{\Delta A}$ be as in (6.3) with density $\eta \leq \frac{1}{2}$;
- $\mathbf{x} \in \mathbb{R}^n$ be as in (m$_1$) with finite $\mathcal{E}_\mathbf{x} = \mathbb{E}[\sum_{j=0}^{n-1} x_j^2]$, $\mathcal{F}_\mathbf{x} = \mathbb{E}[(\sum_{j=0}^{n-1} x_j^2)^2]$.

For any $\theta \in (0,1)$, and any instance of $\mathbf{y} = \mathbf{A}^{(1)}\mathbf{x}$, $\hat{\mathbf{x}}$ that satisfies $\mathbf{y} = \mathbf{A}^{(0)}\hat{\mathbf{x}}$ is such that

$$\mathbb{P}\left[\|\hat{\mathbf{x}} - \mathbf{x}\|_2^2 \geq \frac{4\eta m \mathcal{E}_\mathbf{x}}{\sigma_{\max}(\mathbf{A}^{(0)})^2} \theta\right] \geq \zeta \quad (6.6)$$

where

$$\zeta = \frac{1}{1 + (1-\theta)^{-2}\left[\left[1 + \frac{1}{m}\left(\frac{3}{2\eta} - 1\right)\right]\frac{\mathcal{F}_\mathbf{x}}{\mathcal{E}_\mathbf{x}^2} - 1\right]} \quad (6.7)$$

This is extended to the asymptotic case (*i.e.*, model (m$_2$)) as follows.

Theorem 6.2 *(Second-class recovery error lower bound (asymptotic case))*. Let:

- $\mathbf{A}^{(0)}, \mathbf{A}^{(1)}, \mathbf{\Delta A}, \eta$ be as in Theorem 6.1 as $m, n \to \infty$, $\frac{m}{n} \to q$;
- $\mathcal{X}$ be as in (m$_2$), α-mixing [117, (27.25)], with finite $\mathcal{W}_\mathbf{x} = \lim_{n\to\infty} \frac{1}{n}\mathbb{E}\left[\sum_{j=0}^{n-1} x_j^2\right]$ and uniformly-bounded $\mathbb{E}[X_j^4] \leq m_\mathbf{x}$ for some $m_\mathbf{x} > 0$.

For any $\theta \in (0,1)$, and any instance of $\mathbf{y} = \frac{1}{\sqrt{n}}\mathbf{A}^{(1)}\mathbf{x}$, $\hat{\mathbf{x}}$ that satisfies $\mathbf{y} = \frac{1}{\sqrt{n}}\mathbf{A}^{(0)}\hat{\mathbf{x}}$ is such that[a]

$$\mathbb{P}\left[\mathcal{W}_{\hat{\mathbf{x}}-\mathbf{x}} \geq \frac{4\eta q \mathcal{W}_\mathbf{x}}{(1+\sqrt{q})^2}\theta\right] \simeq 1 \quad (6.8)$$

---

[a] Clearly the recovery error power $\mathcal{W}_{\hat{\mathbf{x}}-\mathbf{x}} = \lim_{n\to\infty} \frac{1}{n}\sum_{j=0}^{n-1}(\hat{x}_j - x_j)^2$.

The proof of these statements is given below. Simply put, Theorems 6.1 and 6.2 state that a second-class decoder recovering with any algorithm $\hat{\mathbf{x}}$ such that $\mathbf{y} = \mathbf{A}^{(0)}\hat{\mathbf{x}}$ is subject to a recovery error whose norm, with high probability, exceeds a quantity depending on the density η of the perturbation $\boldsymbol{\Delta A}$, the undersampling rate m/n and the average energy $\mathcal{E}_\mathbf{x}$ or power $\mathcal{W}_\mathbf{x}$ respectively. In particular, the non-asymptotic case in (6.6) is a probabilistic lower bound: as a quantitative example, by assuming it holds with probability $\zeta = 0.98$ and that $\frac{\mathcal{F}_\mathbf{x}}{\mathcal{E}_\mathbf{x}^2} = 1.0001, n = 1024, m = 512, \sigma_{\max}(\mathbf{A}^{(0)}) \simeq \sqrt{m} + \sqrt{n}$ (*i.e.*, as in Theorem 1.4) one could take an arbitrary $\theta = 0.1 \Rightarrow \eta = 0.1594$ to obtain $\|\hat{\mathbf{x}} - \mathbf{x}\|_2^2 \geq 0.0109$ w.r.t. R.V.s having average energy $\mathcal{E}_\mathbf{x} = 1$. In other words, with probability 0.98 a perturbation of density $\eta = 0.1594$ will cause a minimum recovery error norm of 19.61 dB.

A stronger asymptotic result holding with probability 1 on the recovery error power $\mathcal{W}_{\hat{\mathbf{x}}-\mathbf{x}}$ is then reported in Theorem 6.2 under broadly verified assumptions on the R.P. $\mathcal{X}$, where θ can be arbitrarily close to 1 and only affecting the convergence rate to this lower bound. The bounds in (6.6) and (6.8) are adopted as reference best-cases in absence of other noise sources for the second-class decoder, which actually exhibits higher recovery error for most problem instances and reconstruction algorithms as well illustrated in the exemplary applications of Section 6.3.

Proof of the Second-Class Recovery Error Lower Bound

In this Section we give a technical proof of Theorems 6.1 and 6.2. We first introduce a Lemma that gives a self-contained probabilistic result on the Euclidean norm of ϵ in (6.5).

Lemma 6.1. Let:

- $\xi \in \mathbb{R}^n$ be a R.V. with $\mathcal{E}_\xi = \mathbb{E}[\sum_{j=0}^{n-1} \xi_j^2]$, $\mathcal{F}_\xi = \mathbb{E}[(\sum_{j=0}^{n-1} \xi_j^2)^2]$;
- $\boldsymbol{\Delta A} \in \{-2, 0, 2\}^{m \times n}$ be the sparse random matrix in (6.3) drawn from a RME with i.i.d. entries and density $\eta = \frac{c}{mn} \leq \frac{1}{2}$.

If $\boldsymbol{\xi}$ and $\boldsymbol{\Delta A}$ are independent, then for any $\theta \in (0,1)$

$$\mathbb{P}\left[\|\boldsymbol{\Delta A}\boldsymbol{\xi}\|_2^2 \geq 4mn\,\mathcal{E}_{\boldsymbol{\xi}}\theta\right] \geq \zeta \tag{6.9}$$

with

$$\zeta = \left\{1 + (1-\theta)^{-2}\left[\left(1 + \frac{1}{m}(\frac{3}{2\eta}-1)\right)\frac{\mathcal{F}_{\boldsymbol{\xi}}}{\mathcal{E}_{\boldsymbol{\xi}}^2} - 1\right]\right\}^{-1} \tag{6.10}$$

Proof of Lemma 6.1. Consider

$$\|\boldsymbol{\Delta A}\boldsymbol{\xi}\|_2^2 = \sum_{j=0}^{m-1}\sum_{l=0}^{n-1}\sum_{i=0}^{n-1}\Delta A_{j,l}\Delta A_{j,i}\xi_l\xi_i$$

We now derive the first and second moments of this positive r.v. as follows; $\boldsymbol{\Delta A}$ is drawn from a RME of i.i.d. entries with mean $\mu_{\boldsymbol{\Delta A}} = 0$, variance $\sigma_{\boldsymbol{\Delta A}}^2 = 4\eta$ and $\forall (j,l) \in \{0,\ldots,m-1\} \times \{0,\ldots,n-1\}$, $\mathbb{E}[\Delta A_{j,l}^4] = 16\eta$. Using the independence between $\boldsymbol{\xi}$ and $\boldsymbol{\Delta A}$, and the fact that $\boldsymbol{\Delta A}$ is drawn from a RME with i.i.d. entries we have that the first moment

$$\mathbb{E}\left[\|\boldsymbol{\Delta A}\boldsymbol{\xi}\|_2^2\right] = \sum_{j=0}^{m-1}\sum_{l=0}^{n-1}\sum_{i=0}^{n-1}\mathbb{E}[\Delta A_{j,l}\Delta A_{j,i}]\mathbb{E}[\xi_l\xi_i]$$

$$= \sum_{j=0}^{m-1}\sum_{l=0}^{n-1}\sum_{i=0}^{n-1}\sigma_{\boldsymbol{\Delta A}}^2\delta(l,i)\mathbb{E}[\xi_l\xi_i] = \sum_{j=0}^{m-1}\sigma_{\boldsymbol{\Delta A}}^2\sum_{l=0}^{n-1}\mathbb{E}[\xi_l^2] = 4mn\,\mathcal{E}_{\boldsymbol{\xi}}$$

For the aforementioned properties of $\boldsymbol{\Delta A}$ we also have

$$\mathbb{E}[\Delta A_{j,l}\Delta A_{j,i}\Delta A_{v,h}\Delta A_{v,o}] = \begin{cases} \sigma_{\boldsymbol{\Delta A}}^4, & \begin{cases} j \neq v, l = i, h = o \\ j = v, l = i, h = o, l \neq h \\ j = v, l = h, i = o, l \neq i \\ j = v, l = o, i = h, l \neq i \end{cases} \\ \mathbb{E}[\Delta A_{j,l}^4], & j = v, l = i = h = o \\ 0, & \text{otherwise} \end{cases} \tag{6.11}$$

illustrating the expectation of all possible 4-ples of entries of $\mathbf{\Delta A}$. After cumbersome but straightforward calculations that involve the substitution of (6.11) into $\mathbb{E}\left[(\|\mathbf{\Delta A}\boldsymbol{\xi}\|_2^2)^2\right]$ we obtain

$$\mathbb{E}\left[(\|\mathbf{\Delta A}\boldsymbol{\xi}\|_2^2)^2\right] = 16m\eta(\eta(m-1)\mathcal{F}_{\boldsymbol{\xi}} + 3\eta(\mathcal{F}_{\boldsymbol{\xi}} - \mathcal{G}_{\boldsymbol{\xi}}) + \mathcal{G}_{\boldsymbol{\xi}})$$

where $\mathcal{G}_{\boldsymbol{\xi}} = \mathbb{E}\left[\sum_{j=0}^{n-1} \xi_j^4\right]$. We are now in the position of using a one-sided version of Chebyshev's inequality for positive r.v.s, *i.e.*, any r.v. $z \geq 0$ verifies

$$\forall \theta \in (0,1), \ \mathbb{P}\left[z \geq \theta \mathbb{E}[z]\right] \geq \frac{(1-\theta)^2 \mu_z^2}{(1-\theta)^2 \mu_z^2 + \sigma_z^2} \quad (6.12)$$

By applying this inequality to $\|\mathbf{\Delta A}\boldsymbol{\xi}\|_2^2$ we have that, $\forall \theta \in (0,1)$,

$$\mathbb{P}\left[\|\mathbf{\Delta A}\boldsymbol{\xi}\|_2^2 \geq \theta \mathbb{E}[\|\mathbf{\Delta A}\boldsymbol{\xi}\|_2^2]\right]$$

$$\geq \left[1 + (1-\theta)^{-2}\left[\frac{\mathbb{E}[(\|\mathbf{\Delta A}\boldsymbol{\xi}\|_2^2)^2]}{\mathbb{E}[\|\mathbf{\Delta A}\boldsymbol{\xi}\|_2^2]^2} - 1\right]\right]^{-1}$$

$$= \left[1 + (1-\theta)^{-2}\left[\left(1 - \frac{1}{m}\right)\frac{\mathcal{F}_{\boldsymbol{\xi}}}{\mathcal{E}_{\boldsymbol{\xi}}^2} + \frac{3\eta(\mathcal{F}_{\boldsymbol{\xi}} - \mathcal{G}_{\boldsymbol{\xi}}) + \mathcal{G}_{\boldsymbol{\xi}}}{\eta m \mathcal{E}_{\boldsymbol{\xi}}^2} - 1\right]\right]^{-1}$$

which yields (6.10) by considering that when $\eta \leq \frac{1}{2}$, $3\eta(\mathcal{F}_{\boldsymbol{\xi}} - \mathcal{G}_{\boldsymbol{\xi}}) + \mathcal{G}_{\boldsymbol{\xi}} \leq \frac{3}{2}\mathcal{F}_{\boldsymbol{\xi}}$. □

We are now in the position of proving Theorem 6.1.

Proof of Theorem 6.1. Since all decoders receive in absence of other noise sources the same measurements $\mathbf{y} = \mathbf{A}^{(1)}\mathbf{x}$, a second-class decoder would naively assume $\mathbf{y} = \mathbf{A}^{(0)}\hat{\mathbf{x}}$ with $\hat{\mathbf{x}}$ an approximation of $\mathbf{x}$ obtained by a decoder that satisfies this equality, *e.g.*, as the naive BP in Section 5.1.2. Since $\mathbf{A}^{(1)} = \mathbf{A}^{(0)} + \mathbf{\Delta A}$, if we define $\mathbf{\Delta x} = \hat{\mathbf{x}} - \mathbf{x}$ we may write $\mathbf{A}^{(0)}\mathbf{x} + \mathbf{\Delta A}\mathbf{x} = \mathbf{A}^{(0)}\hat{\mathbf{x}}$ and thus $\mathbf{A}^{(0)}\mathbf{\Delta x} = \mathbf{\Delta A}\mathbf{x}$. $\|\mathbf{\Delta x}\|_2^2$ can then be bounded straightforwardly as $\sigma_{\max}(\mathbf{A}^{(0)})^2 \|\mathbf{\Delta x}\|_2^2 \geq \|\mathbf{\Delta A}\mathbf{x}\|_2^2$ yielding

$$\|\hat{\mathbf{x}} - \mathbf{x}\|_2^2 \geq \frac{\|\mathbf{\Delta A}\mathbf{x}\|_2^2}{\sigma_{\max}(\mathbf{A}^{(0)})^2} \quad (6.13)$$

By applying the probabilistic lower bound of Lemma 6.1 on $\|\mathbf{\Delta A}\mathbf{x}\|_2^2$ in (6.13), we have that $\|\mathbf{\Delta A}\mathbf{x}\|_2^2 \geq 4m\eta \mathcal{E}_{\mathbf{x}} \theta$ for $\theta \in (0,1)$ and a given probability value exceeding ζ in (6.10). Plugging the *right-hand side (RHS)* of this inequality in (6.13) yields (6.6). □

The following Lemma applies to finding the asymptotic result (6.8) of Theorem 6.2.

Lemma 6.2. *Let $\mathcal{X}$ be an α-mixing R.P. with uniformly-bounded fourth moments $\mathbb{E}[x_j^4] \leq m_\mathbf{x}$ for some $m_\mathbf{x} > 0$. Define*

$$\mathcal{E}_\mathbf{x} = \mathbb{E}\left[\sum_{j=0}^{n-1} x_j^2\right]$$

and

$$\mathcal{F}_\mathbf{x} = \mathbb{E}\left[\left(\sum_{j=0}^{n-1} x_j^2\right)^2\right]$$

If

$$\mathcal{W}_\mathbf{x} = \lim_{n\to\infty} \frac{1}{n}\mathcal{E}_\mathbf{x} > 0$$

then

$$\lim_{n\to\infty} \frac{\mathcal{F}_\mathbf{x}}{\mathcal{E}_\mathbf{x}^2} = 1$$

Proof of Lemma 6.2. Note first that from Jensen's inequality $\mathcal{F}_\mathbf{x} \geq \mathcal{E}_\mathbf{x}^2$, so $\lim_{n\to\infty} \frac{1}{n}\mathcal{E}_\mathbf{x} > 0$ also implies that $\lim_{n\to\infty} \frac{1}{n^2}\mathcal{E}_\mathbf{x}^2 > 0$ and $\lim_{n\to\infty} \frac{1}{n^2}\mathcal{F}_\mathbf{x} > 0$. Since $\lim_{n\to\infty} \frac{1}{n^2}\mathcal{E}_\mathbf{x}^2 = \mathcal{W}_\mathbf{x}^2 > 0$ we may write

$$\lim_{n\to\infty} \frac{\mathcal{F}_\mathbf{x}}{\mathcal{E}_\mathbf{x}^2} = 1 + \frac{\lim_{n\to\infty} \frac{1}{n^2}\mathcal{F}_\mathbf{x} - \frac{1}{n^2}\mathcal{E}_\mathbf{x}^2}{\mathcal{W}_\mathbf{x}^2} \qquad (6.14)$$

and observe that

$$\left|\frac{1}{n^2}\mathcal{F}_\mathbf{x} - \frac{1}{n^2}\mathcal{E}_\mathbf{x}^2\right| \leq \frac{1}{n^2}\sum_{j=0}^{n-1}\sum_{l=0}^{n-1}|\Xi_{j,l}|$$

where

$$\Xi_{j,l} = \mathbb{E}[x_j^2 x_l^2] - \mathbb{E}[x_j^2]\mathbb{E}[x_l^2] = \mathbb{E}[(x_j^2 - \mathbb{E}[x_j^2])(x_l^2 - \mathbb{E}[x_l^2])]$$

From the α-mixing assumption we know that $|\Xi_{j,l}| \leq \alpha(|j-l|) \leq m_\mathbf{x}$ with the sequence $\alpha(h)$ vanishing to 0 as $h \to \infty$. Hence,

$$\left|\frac{1}{n^2}\mathcal{F}_\mathbf{x} - \frac{1}{n^2}\mathcal{E}_\mathbf{x}^2\right| \leq \frac{1}{n^2}\sum_{j=0}^{n-1}|\Xi_{j,j}| + \frac{2}{n^2}\sum_{h=1}^{n-1}\sum_{j=0}^{n-h-1}|\Xi_{j,j+h}|$$

$$\leq \frac{n\,m_\mathbf{x}}{n^2} + \frac{2}{n^2}\sum_{h=1}^{n-1}(n-h)\alpha(h) \leq \frac{m_\mathbf{x}}{n} + \frac{2}{n}\sum_{h=1}^{n-1}\alpha(h) \quad (6.15)$$

The thesis of this Lemma follows from the fact that the upper bound in (6.15) vanishes to 0 as $n \to \infty$. This is obvious when $\sum_{h=0}^{+\infty}\alpha(h)$ is convergent. Otherwise, if $\sum_{h=0}^{+\infty}\alpha(h)$ is divergent we may resort to the Stolz-Cesàro theorem to find $\lim_{n\to\infty}\frac{1}{n}\sum_{h=1}^{n-1}\alpha(h) = \lim_{n\to\infty}\alpha(n) = 0$. □

We are now in the position of proving Theorem 6.2, that is a mere extension of the proof of Theorem 6.1 to the asymptotic case.

Proof of Theorem 6.2. The inequality (6.13) in the proof of Theorem 6.1 is now modified for the asymptotic case, *i.e.*, for a R.P. $\mathcal{X}$. Note that $\mathbf{A}^{(0)}$ is drawn from the RBE with zero mean, unit variance entries; thus, when $m, n \to \infty$ with $m/n \to q$ the value $\sqrt{n}\,\sigma_{\max}(\mathbf{A}^{(0)})$ is known from [83] (*i.e.*, as in Theorem 1.4) since all its singular values belong to the interval $\left[1 - \frac{1}{\sqrt{q}}, 1 + \frac{1}{\sqrt{q}}\right]$. We therefore assume $\sigma_{\max}(\mathbf{A}^{(0)}) \simeq \sqrt{m} + \sqrt{n}$ and take the limit of (6.13) normalised by $1/n$ for $m, n \to \infty$, *i.e.*, the recovery error power

$$W_{\hat{\mathbf{x}}-\mathbf{x}} = \lim_{n\to\infty}\frac{1}{n}\sum_{j=0}^{n-1}(\hat{x}_j - x_j)^2 \geq \lim_{m,n\to\infty}\frac{\left\|\Delta\mathbf{A}\frac{\mathbf{x}^{(n)}}{\sqrt{n}}\right\|_2^2}{(\sqrt{m}+\sqrt{n})^2} \quad (6.16)$$

with $\mathbf{x}^{(n)}$ the n-th finite-length term in a plaintext $\mathbf{x} = \{\mathbf{x}^{(n)}\}_{n=0}^{+\infty}$ of $\mathcal{X}$. We may now apply Lemma 6.1 in $\boldsymbol{\xi} = \frac{\mathbf{x}^{(n)}}{\sqrt{n}}$ for each $\|\Delta\mathbf{A}\boldsymbol{\xi}\|_2^2$ at the numerator of the RHS of (6.16) with $\mathcal{F}_{\boldsymbol{\xi}} = \frac{1}{n^2}\mathcal{F}_\mathbf{x}$, $\mathcal{E}_{\boldsymbol{\xi}} = \frac{1}{n}\mathcal{E}_\mathbf{x}$ and $\mathcal{E}_\mathbf{x}, \mathcal{F}_\mathbf{x}$ as in Lemma 6.2. For $m, n \to \infty$ and $\eta \leq \frac{1}{2}$, the probability in (6.10) becomes

$$\forall \theta \in (0,1), \lim_{m,n\to\infty}\zeta = \left[1 + (1-\theta)^{-2}\left[\lim_{n\to\infty}\frac{\frac{1}{n^2}\mathcal{F}_\mathbf{x}}{\frac{1}{n^2}\mathcal{E}_\mathbf{x}^2} - 1\right]\right]^{-1}$$

Since $\mathcal{X}$ also satisfies by hypothesis the assumptions of Lemma 6.2, we have that
$$\lim_{n\to\infty} \frac{\mathcal{F}_\xi}{\mathcal{E}_\xi^2} = 1$$
and thus $\lim_{m,n\to\infty} \zeta = 1$. Hence, with $m/n \to q$ and probability 1 the RHS of (6.16) becomes
$$\forall \theta \in (0,1), \quad \lim_{m,n\to\infty} \frac{\|\Delta\mathbf{A}\xi\|_2^2}{n(1+\sqrt{\frac{m}{n}})^2} = \lim_{m,n\to\infty} \frac{4m\eta\mathcal{E}_\mathbf{x}}{n^2(1+\sqrt{\frac{m}{n}})^2}\theta$$
and the recovery error power is shown to satisfy (6.8). □

Thus, Theorems 6.1 and 6.2 were shown to hold in the respective cases.

6.2.2 Second-Class Recovery Error Norm: Upper Bound

The second-class recovery error norm is substantially bounded from above by direct application of Theorem 5.1. To apply it, we have to compute $\epsilon_{\mathbf{A}^{(1)}}^{(k)}, \epsilon_{\mathbf{A}^{(1)}}^{(2k)}$ in our particular case. Theoretical results exist for estimating their value by bounding the extreme singular values in (5.1), since both $\mathbf{A}^{(1)}$ and $\Delta\mathbf{A}$ are drawn from i.i.d. RMEs.

To estimate $\epsilon_{\mathbf{A}^{(1)}}^{(k)}$ we may proceed in the following fashion: since $\Delta\mathbf{A}$ is drawn from a RME with i.i.d. zero-mean entries for which

$$\forall (j,l) \in \{0,\ldots,m-1\} \times \{0,\ldots,n-1\},$$
$$\mathbb{E}[\Delta A_{j,l}^2] = 4\eta, \mathbb{E}[\Delta A_{j,l}^4] = 16\eta$$

we may use [169, Theorem 2] to find

$$\mathbb{E}[\sigma_{\max}^{(k)}(\Delta\mathbf{A})] = 2\underline{c}'(\sqrt{k\eta} + \sqrt{m\eta} + (mk\eta)^{\frac{1}{4}}) \quad (6.17)$$

for $\underline{c}' > 0$ a universal constant. Then, using the non-asymptotic estimate of Theorem 1.3, we may assume $\sigma_{\max}^{(k)}(\mathbf{A}^{(1)}) = \underline{c}''(\sqrt{k} + \sqrt{m})$ for another universal constant $\underline{c}'' > 0$. Thus, we have

$$\epsilon_{\mathbf{A}^{(1)}}^{(k)} \simeq 2\underline{c}\frac{\sqrt{k\eta} + \sqrt{m\eta} + (mk\eta)^{\frac{1}{4}}}{\sqrt{k} + \sqrt{m}} \quad (6.18)$$

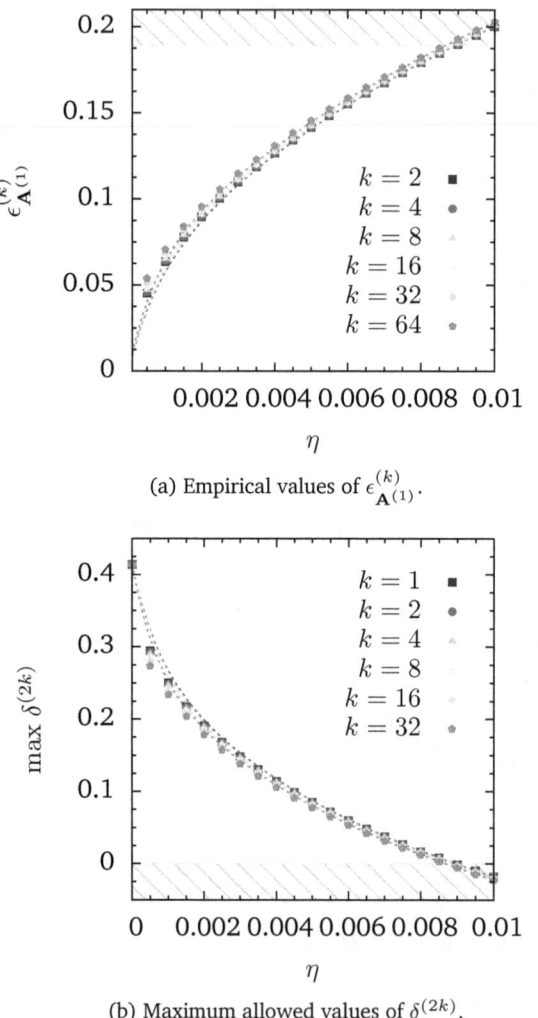

(a) Empirical values of $\epsilon_{\mathbf{A}^{(1)}}^{(k)}$.

(b) Maximum allowed values of $\delta^{(2k)}$.

Figure 6.4: Empirical evaluation of the constants in Theorem 5.1 based on a large number of $\mathbf{A}^{(1)}$, $\mathbf{\Delta A}$ with $m = 512, \eta \in [5 \cdot 10^{-4}, 10^{-2}]$ and $\mathbf{D}$ a random ONB. The forbidden areas in the statement of Theorem 5.1 are marked with stripes.

for $\underline{c} = \frac{c'}{c''} > 0$ a universal constant, where the approximation is due to the fact that (6.17) actually yields an expectation of the maximum. However, this estimate is easily applicable only when $\mathbf{D}$ is the canonical basis; since in many practical cases this does not hold, we simply resort to a Monte Carlo simulation of $\epsilon^{(k)}_{\mathbf{A}^{(1)}}$ with, e.g., $\mathbf{D}$ a random ONB. As an example of such a numerical analysis, we calculate (5.1) for 10^4 instances of submatrices of $\mathbf{A}^{(1)}$ and $\mathbf{\Delta A}$ with $m = 512, k = 2, 4, \ldots, 64$ and $\eta \in [5 \cdot 10^{-4}, 10^{-2}]$. This allows us to find typical values of $\epsilon^{(k)}_{\mathbf{A}^{(1)}}$ as reported in Fig. 6.4a. In this test case, we have found that $\underline{c} \approx 0.5741$ in (6.18) would match the simulations. In the same setting $\epsilon^{(k)}_{\mathbf{A}^{(1)}} < 2^{\frac{1}{4}} - 1$ only when $\eta \leq 8 \cdot 10^{-3}$. In Fig. 6.4b we report the corresponding range of allowed constants $\delta^{(2k)} \leq \delta^{(2k)}_{\max}$ that comply with Theorem 5.1, i.e., the RIP constraints the encoding matrices must meet so that (5.2) holds.

Once again, RIP-based analyses provide very strong sufficient conditions for signal recovery, which in our case result in establishing a formal upper bound for a small range of η. As observed by the very authors of [91], typical recovery errors are substantially smaller than this upper bound. We will therefore rely on another less rigorous, yet practically effective least-squares approach using the same hypotheses of Theorem 6.1 to bound the average recovery quality performances, as presented in the following Section.

6.2.3 Average Signal-to-Noise Ratio Bounds

We have already discussed how the perturbation density η is the main design parameter for the proposed multiclass encryption by CS, and have presented in Chapter 5 a method to predict the average recovery performances under a variety of perturbations, including the SSF which is at the heart of our encryption scheme. To provide *criteria* for the choice of η we adopt two $\text{ASNR}_{\hat{x}, x}$ bounds derived as follows.

The Lower Bound

Although rigorous, the second- (or lower-) class recovery error upper bound derived by applying Theorem 5.1 is only compatible with small values of (k, η), as shown by the evidence gathered in Fig. 6.4. To bound the typical recovery performances in a larger range we follow a method similar to the one used in Section 5.2, *i.e.*, we analyse the behaviour of a lower-class decoder that naively recovers $\hat{x}$ such that $y = A^{(0)}\hat{x} = (A^{(0)} + \Delta A)x$ and $A^{(0)}(\hat{x} - x) = \Delta A x$. In most cases, such a recovery produces $\hat{x}$ lying close to x, so we approximate $\hat{x} - x = (A^{(0)})^\dagger \Delta A x$, *i.e.*,

$$\frac{\|\hat{x} - x\|_2^2}{\|x\|_2^2} \leq \sigma_{\max}((A^{(0)})^\dagger \Delta A)^2$$

By taking an empirical expectation on both sides, our criterion becomes $\mathrm{ASNR}_{\hat{x},x} > \mathrm{LB}(m, n, \eta)$ where

$$\mathrm{LB}(m, n, \eta) = -10 \log_{10} \hat{\mathbb{E}}\left(\sigma_{\max}((A^{(0)})^\dagger \Delta A)^2\right) \mathrm{dB} \quad (6.19)$$

(6.19) is then calculated by a thorough Monte Carlo simulation of $\sigma_{\max}((A^{(0)})^\dagger \Delta A)$.

The Upper Bound

The opposite criterion is found by assuming $\mathrm{ASNR}_{\hat{x},x} < \mathrm{UB}(m, n, \eta)$ where

$$\mathrm{UB}(m, n, \eta) = -10 \log_{10} \frac{4\eta m}{(\sqrt{m} + \sqrt{n})^2} \mathrm{dB} \quad (6.20)$$

that is obtained from a simple rearrangement of (6.8) with $\theta \simeq 1$. We will see how (6.19) and (6.20) fit well the $\mathrm{ASNR}_{\hat{x},x}$ performances of the examples, and enable a sufficiently reliable estimate of the range of performances of lower-class receivers from a given configuration of (m, n, η).

6.3 Performance Evaluation

In this section we detail some example applications for the multiclass CS scheme we proposed. For each exemplary case, we study the recovery quality attained by first-class receivers against second-class ones in the two-class scheme for the sake of simplicity; these results encompass the multiclass setting since high-class receivers will correspond to first-class recovery performances (*i.e.*, at a perturbation density $\eta = 0$), while lower-class users will attain the performances of a second-class receiver at a fixed $\eta > 0$.

6.3.1 Experimental Framework

For each plaintext $\mathbf{x} = \mathbf{D}\mathbf{s}$ being reconstructed and each approximation $\hat{\mathbf{x}} = \mathbf{D}\hat{\mathbf{s}}$, we evaluate once again the $\mathrm{ASNR}_{\hat{\mathbf{x}},\mathbf{x}}$ of (1.32); this average performance index is compared against (6.19) and (6.20) with the purpose of choosing a suitable perturbation density η so that lower-class recovery performances are set to the desired quality level. In particular, each example reports (6.19) obtained by a Monte Carlo simulation of the singular values of $(\mathbf{A}^{(0)})^{\dagger}\mathbf{\Delta A}$ over $5 \cdot 10^3$ cases.

Since our emphasis is on showing that, despite its simplicity, our method is effective in avoiding the access to high-quality information content for lower-class receivers, we complement the $\mathrm{ASNR}_{\hat{\mathbf{x}},\mathbf{x}}$ evidence of each example with an automated assessment of the information content intelligible from $\hat{\mathbf{x}}$ by means of feature-extraction algorithms. These are equivalent to partially informed attacks to the encryption, attempting to expose the sensitive content inferred from the recovered signal. More specifically, we will try to recover an English sentence from a speech segment, the location of the PQRST peaks in an ECG, and printed text in an image.

6.3.2 Recovery Algorithms

While we have widely discussed the use of BP and BPDN in this thesis, and in particular w.r.t. their sensitivity to matrix perturbations in Section 5.1. These convex problems are often replaced in practice by a variety of high-performance algorithms, and in detail probabilistic inference algorithms such as those in [104, 105] are capable of solving BPDN with statistical priors on the nature of noise; thus, they are particularly well-fit to our application if we want to assess the best achievable performances of lower-class decoders. For completeness, as reference cases for most common algorithmic classes we preliminarily tested the solution of BPDN as implemented in SPGL_1; this was compared to the greedy algorithm *Compressive Sampling Matching Pursuit (CoSaMP)* [99] and the *Generalised Approximate Message-Passing (GAMP)* algorithm [105].

To optimise these preliminary tests, the algorithms were optimally tuned in a "genie" fashion: BPDN was solved as $\Delta_{\rm BPDN}(\mathbf{y}, \mathbf{A}^{(0)}\mathbf{D}, \varepsilon^*)$, *i.e.*, as if $\varepsilon^* = \|\boldsymbol{\Delta}\mathbf{A}\mathbf{x}\|_2$ was known beforehand; CoSaMP was initialised with the exact sparsity level k for each case; GAMP was run with the sparsity-enforcing, i.i.d. Bernoulli-Gaussian prior (see, *e.g.*, [170]) and initialised with the exact sparsity ratio k/n of each instance, and the exact mean and variance of each considered test set. Moreover, signal-independent parameters were hand-tuned in each case to yield optimal recovery performances.

For the sake of brevity, in each example we select and report the algorithm that yields the most accurate recovery quality at a lower-class decoder as the amount of perturbation varies. We found that GAMP achieves the highest $\mathrm{ASNR}_{\hat{\mathbf{x}},\mathbf{x}}$ in all the settings explored in the examples, consistently with the observations in [170] that assess the robust recovery capabilities of this algorithm under a broadly applicable sparsity-enforcing prior. Moreover, as $\boldsymbol{\Delta}\mathbf{A}$ verifies [160, Proposition 2.1] the perturbation noise ϵ is approximately Gaussian for large (m, n) and thus GAMP tuned as above yields the optimal performances as expected.

Note that recovery algorithms which attempt to jointly identify $\mathbf{x}$ and $\boldsymbol{\Delta}\mathbf{A}$ [159, 160] can be seen as explicit attacks to multiclass encryption and are thus evaluated in Chapter 7, anticipating that their performances are compatible with those of GAMP.

6.3.3 Speech Signals

We consider a subset of spoken English sentences from the PTDB-TUG database [171] with original sampling frequency $f_s = 48\,\mathrm{kHz}$, variable duration and sentence length. Each speech signal is divided in segments of $n = 512$ samples and encoded by two-class CS with $m = \frac{n}{2}$ measurements. We obtain the sparsity basis $\mathbf{D}$ by applying principal component analysis to 500 n-dimensional segments yielding

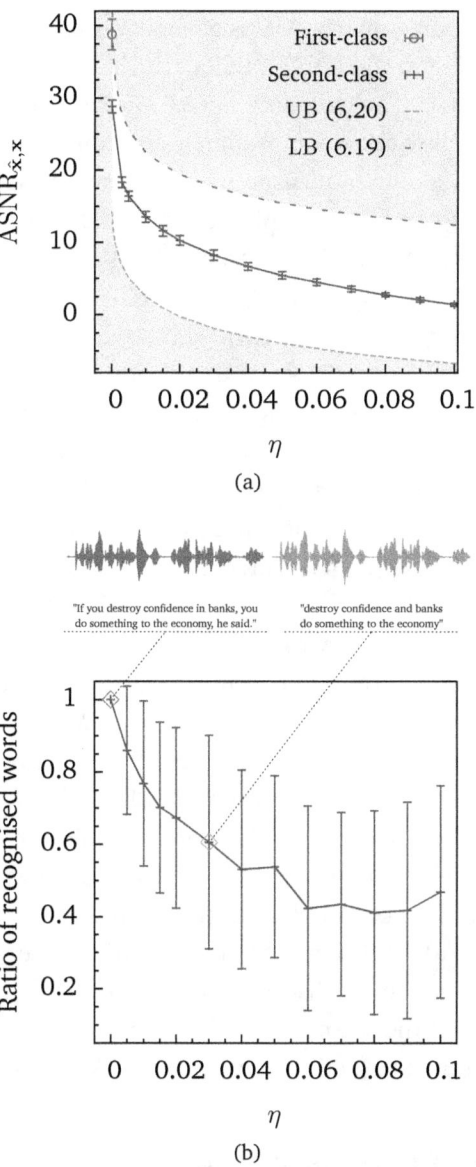

Figure 6.5: Multiclass CS of speech signals: (a) $\mathrm{ASNR}_{\hat{\mathbf{x}},\mathbf{x}}$ as a function of the perturbation density $\eta \in [0, 0.1]$ (solid) and second-class $\mathrm{SNR}_{\hat{\mathbf{x}},\mathbf{x}}$ upper bound (dashed); (b) Ratio of words correctly recognised by ASR in $\eta \in [0, 0.1]$ (bottom) and typical recovered instances for $\eta = \{0, 0.03\}$ (top).

an ONB. The encoding matrix $\mathbf{A}^{(1)}$ is generated from $\mathbf{A}^{(0)}$ drawn from the RBE, by adding to the latter a SSF perturbation $\Delta \mathbf{A}$ chosen as in (6.3) with density η. The encoding in (6.5) is simulated in a realistic setting, where each window $\mathbf{x}$ of n samples is acquired with a different instance of $\mathbf{A}^{(1)}$ yielding m measurements per speech segment. As for the decoding stage, we apply GAMP as specified above to recover $\hat{\mathbf{x}}$ given $\mathbf{A}^{(1)}$ (first-class) and $\mathbf{A}^{(0)}$ (second-class).

For a given encoding matrix a first-class receiver is capable of decoding a clean speech signal with $\mathrm{ASNR}_{\hat{\mathbf{x}},\mathbf{x}} = 38.76\,\mathrm{dB}$, whereas a second-class receiver is subject to significant $\mathrm{ASNR}_{\hat{\mathbf{x}},\mathbf{x}}$ degradation when η increases, as shown in Fig. 6.5a. Note that while the $\mathrm{SNR}_{\hat{\mathbf{x}},\mathbf{x}}$ for $\eta = 0$ has a relative deviation of $2.14\,\mathrm{dB}$ around its mean (*i.e.*, the $\mathrm{ASNR}_{\hat{\mathbf{x}},\mathbf{x}}$), as η increases the observed relative deviation is less than $0.72\,\mathrm{dB}$ due to the perturbation becoming the dominant effect in limiting the recovery quality w.r.t. the fact that $\mathbf{x}$ are compressible by Definition 1.4, but not k-sparse. Note how the $\mathrm{ASNR}_{\hat{\mathbf{x}},\mathbf{x}}$ values lie in the highlighted range between (6.19), (6.20).

To further quantify the limited quality of attained recoveries, we process the recovered signal with the Google Web Speech interface [172, 173] which provides basic *Automatic Speech Recognition (ASR)*. The ratio of words correctly recognised by ASR for different values of η is reported in Fig. 6.5b; there we also depict a typical recovered signal instance, on which a first-class user (*i.e.*, $\eta = 0$) attains $\mathrm{SNR}_{\hat{\mathbf{x}},\mathbf{x}} = 36.58\,\mathrm{dB}$, whereas a second-class decoder only achieves $\mathrm{SNR}_{\hat{\mathbf{x}},\mathbf{x}} = 8.42\,\mathrm{dB}$ when $\eta = 0.03$. The corresponding ratio of recognised words is $14/14$ against $8/14$. In both cases the sentence is intelligible to a human listener, yet the second-class decoder recovers a signal that is sufficiently corrupted to avoid straightforward ASR.

6.3.4 Electrocardiographic Signals

We now process a large subset of ECGs from the PhysioNet database [120] sampled at $f_s = 256\,\mathrm{Hz}$. In particular, we report the case of a typical 25-minutes ECG (sequence e0108) and encode windows of $n = 256$ samples by two-class CS with $m = 90$ measurements,

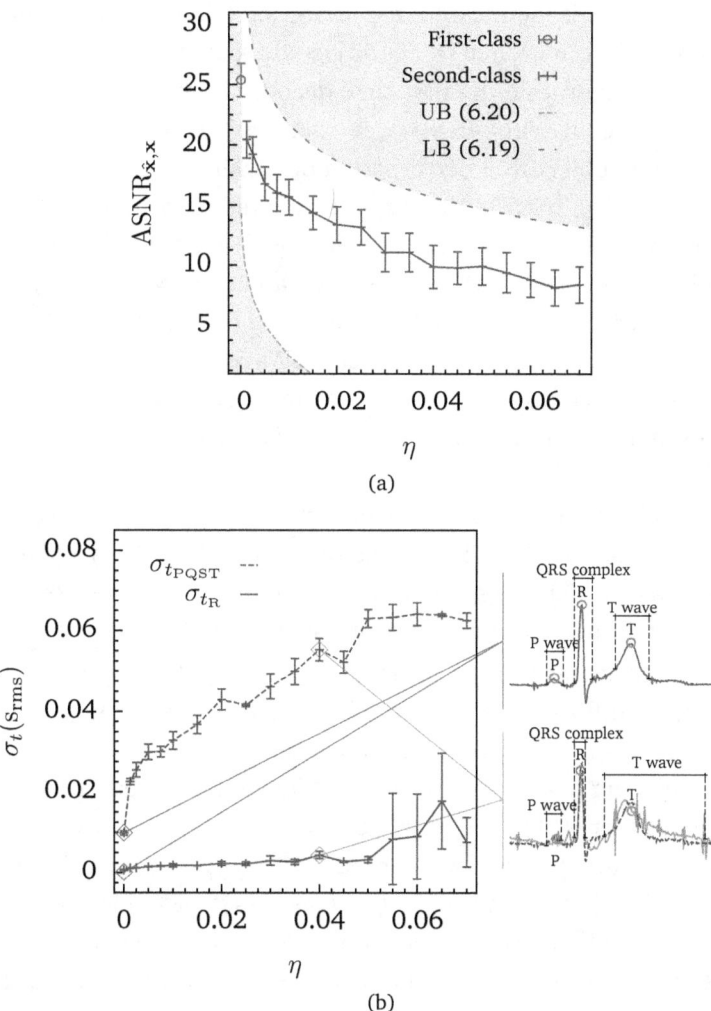

Figure 6.6: Multiclass CS of ECG signals: (a) $\mathrm{ASNR}_{\hat{x},x}$ as a function of the perturbation density $\eta \in [0, 0.05]$ (solid) and second-class $\mathrm{SNR}_{\hat{x},x}$ upper bound (dashed); (b) Time displacement (left) of the R (solid) and P,Q,S,T (dashed) peaks as evaluated by APD for $\eta \in [0, 0.05]$ with typical recovered instances (right) for first-class (top) and second-class (bottom) users.

amounting to a dataset of 1500 ECG instances. The encoding and decoding scheme is identical to that of Section 6.3.3, and we assume the Symmlet-6 orthonormal DWT [52] as the sparsity basis **D**.

In this configuration the first-class decoder is able to reconstruct the original signal with $\text{ASNR}_{\hat{x},x} = 25.36\,\text{dB}$, whereas a second-class decoder subject to a perturbation of density $\eta = 0.03$ achieves an $\text{ASNR}_{\hat{x},x} = 11.08\,\text{dB}$; the recovery degradation depends on η as reported in Fig. 6.6a. As an additional quantification of the encryption at second-class decoders we apply PUWave [174], an *Automatic Peak Detection (APD)* algorithm, to first- and second-class signal reconstructions. In more detail, PUWave is used to detect the position of the P,Q,R,S and T peaks, *i.e.*, the sequence of pulses whose positions and amplitudes summarise the diagnostic properties of an ECG.

The application of this APD yields the estimated peak instants $\hat{t}_{P,Q,R,S,T}$ for each of $J = 1500$ reconstructed signal windows and each decoder class, which are afterwards compared to the corresponding peak instants as detected on the original signal prior to encoding. Thus, we define the average time displacement $\sigma_t = \sqrt{\frac{1}{J}\sum_{i=0}^{J-1}(\hat{t}^{(i)} - t^{(i)})^2}$ and evaluate it for t_R and t_{PQST}. A first-class receiver is subject to a displacement $\sigma_{t_R} = 0.6\,\text{ms}_{\text{rms}}$ of the R-peak and $\sigma_{t_{PQST}} = 9.8\,\text{ms}_{\text{rms}}$ of the remaining peaks w.r.t. the original signal. On the other hand, a second-class user is able to determine the R-peak with $\sigma_{t_R} = 4.4\,\text{ms}_{\text{rms}}$ while the displacement of the other peaks is $\sigma_{t_{PQST}} = 55.3\,\text{ms}_{\text{rms}}$. As η varies in $[0, 0.05]$ this displacement increases as depicted in Fig. 6.6b, thus confirming that a second-class user will not be able to accurately determine the position and amplitude of the peaks with the exception of the R-peak.

6.3.5 Sensitive Text in Images

In this final example we consider an image dataset of people holding printed identification text and apply multiclass CS to selectively hide this sensitive content to lower-class users. The 640×512 pixel images are encoded by CS in 10×8 blocks each of 64×64 pixel while the two-class strategy is only applied to a relevant image area of 3×4 blocks. We adopt as sparsity basis the 2D Daubechies-4 orthonormal DWT [52] and encode each block of $n = 4096$ pixels with $m = 2048$ measurements; two-class encoding is then applied with a SSF perturbation density $\eta \in [0, 0.4]$.

The $\mathrm{ASNR}_{\hat{x},x}$ performances of this example are reported in Fig. 6.7a as averaged on 20 instances per case, showing a rapid degradation of the $\mathrm{ASNR}_{\hat{x},x}$ as η is increased. This degradation is highlighted in the typical case of Fig. 6.7b for $\eta \in \{0.03, 0.2\}$.

In order to assess the effect of our encryption method with an automatic information extraction algorithm, we have applied Tesseract [175], an *Optical Character Recognition (OCR)* algorithm, to the images reconstructed by a second-class user. The text portion in the recovered image data is preprocessed to enhance their quality prior to OCR: the images are first rotated, then we apply standard median filtering to reduce the highpass noise components. Finally, contrast adjustment and thresholding yield the two-level image which is processed by Tesseract. To assess the attained OCR quality we have measured the average number of *Consecutively Recognised Characters (CRC)* from the decoded text image. In Fig. 6.7b the average CRC is reported as a function of η: as the perturbation density increases the OCR fails to recognise an increasing number of ordered characters, *i.e.*, a second-class user progressively fails to extract text content from the decoded image.

Summary

- ▶ Although not perfectly secure, the extremely simple encoding process entailed by CS yields some encryption capabilities with no additional computational complexity, thus providing a limited but zero-cost form of encryption which might be of interest in the design of secure yet resource-limited sensing interfaces.

- ▶ The linear random encoding provided by RBE was modified to envision a multiclass encryption scheme in which all receivers

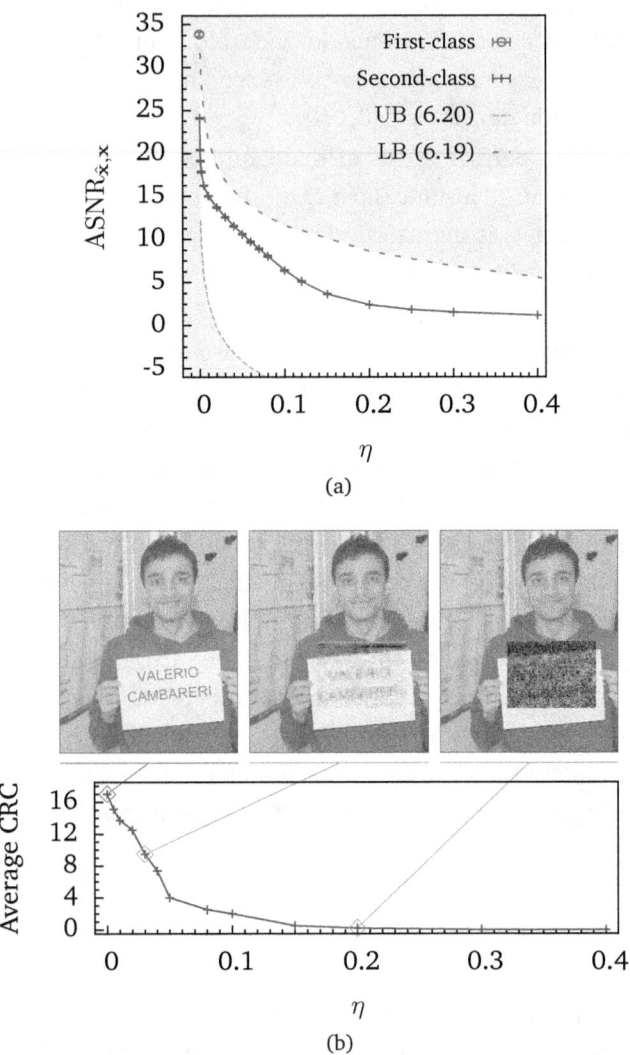

Figure 6.7: Multiclass CS of images: (a) $\mathrm{ASNR}_{\hat{x},x}$ as a function of the perturbation density $\eta \in [0, 0.4]$ (solid) and second-class $\mathrm{SNR}_{\hat{x},x}$ upper bound (dashed); (b) Average CRC by OCR for $\eta \in [0, 0.4]$ (bottom) and typical recovered instances for $\eta \in \{0, 0.03, 0.2\}$ (top).

are given the same set of measurements, but are only enabled to reconstruct the original signal with a decoding quality that depends on their class, *i.e.*, on the private key they possess. This additional design option amounts to the ability of applying SSF to the elements of the encoding matrix, and thus represents an appealing alternative to balance the trade-off between the security of the encoded signal and the resources required to provide it.

▶ The capabilities of multiclass CS were exemplified by simulating the acquisition of sources such as speech segments, ECGs and images with the additional security provided by the devised encryption method.

A Statistical Cryptanalysis of Compressed Sensing

WE now focus on a generic CS configuration $y = Ax$ in the perspective of Section 6.1.1 that linearly encodes a plaintext x into a ciphertext y, and investigate from a statistical perspective the security properties and limits of such linear measurements by letting x, y be realisations of either R.V.s or R.P.s (*i.e.*, (m_1) and (m_2) of Section 6.2) with their respective *a priori* distributions as in the classic Shannon definition of secrecy [176]. The aim of this Chapter is to provide insight on the *achievable* security properties that are granted by a scheme as simple as CS with RBE encoding matrices; this is carried out by evaluating separately the security limits and properties in asymptotic and non-asymptotic configurations of CS.

7.1 Security Limits

The encoding performed by CS is a linear mapping, and as such it cannot completely hide the information contained in a plaintext x. This has two main consequences. Firstly, linearity propagates scaling; hence, it is simple to distinguish a plaintext x' from another x'' if it is known in advance that $x'' = \alpha x'$ for a scalar α. For the particular choice of $\alpha = 0$ this leads to a known argument of Rachlin *et al.* [26, Lemma 1] against the fundamental requirement for secrecy in the Shannon sense.

Proposition 7.1 (*Non-Perfect Secrecy of CS [26]*). Let $\mathbf{x} \in \mathbb{R}^n, \mathbf{y} = \mathbf{A}\mathbf{x} \in \mathbb{R}^m$ be R.V.s representing a plaintext and a ciphertext, $\mathbf{A} \in \mathbb{R}^{m \times n}, m < n$ be any random or deterministic encoding matrix. CS does not have perfect secrecy, *i.e.*, the PDF of the ciphertext conditioned on the plaintext, $f(\mathbf{y}|\mathbf{x}) \neq f(\mathbf{y})$.

To provide insight on the main security limit of CS, we also conveniently report the proof given in [26] with a very slight variation in notation.

Proof of Proposition 7.1. Assume there exists at least one plaintext $\mathbf{x}' \notin \mathrm{Ker}\,(\mathbf{A}) : f(\mathbf{x} = \mathbf{x}') > 0$. Consider the ciphertext $\mathbf{y} = \mathbf{0}_m$; then

$$f(\mathbf{y} = \mathbf{0}_m) = \int_{\mathrm{Ker}(\mathbf{A}) \subseteq \mathbb{R}^n} f(\mathbf{y} = \mathbf{0}_m|\mathbf{x}) f(\mathbf{x}) \mathrm{d}\mathbf{x} = \int_{\mathrm{Ker}(\mathbf{A}) \subseteq \mathbb{R}^n} f(\mathbf{x}) \mathrm{d}\mathbf{x} < 1$$

and for any plaintext $\mathbf{x}'' \in \mathrm{Ker}\,(\mathbf{A})$,

$$f(\mathbf{y} = \mathbf{0}_m|\mathbf{x}'') = 1 \Rightarrow f(\mathbf{y} = \mathbf{0}_m|\mathbf{x}'') \neq f(\mathbf{y} = \mathbf{0}_m)$$

Note how this proof simply relies on the existence of $\mathrm{Ker}\,(\mathbf{A})$ (or equivalently of $\mathrm{Ker}\,(\mathbf{W})$). □

However, will show with a slightly different argument that the information leaking into the ciphertext by means of a linear encoding is the energy of the plaintext, as was partly confirmed in [1, 26]. Moreover, we will prove that with any i.i.d. RsGE encoding matrix a scaling factor α is actually *all* that can be inferred from the statistical analysis of CS-encoded ciphertexts.

Secondly, linearity implies continuity. Hence, whenever $\mathbf{x}'$ and $\mathbf{x}''$ are close to each other for a fixed $\mathbf{A}$, the corresponding $\mathbf{y}'$ and $\mathbf{y}''$ will also be close to each other. This fact goes against the analog version of the *diffusion* (or *avalanche effect*) requirement for digital-to-digital ciphers [176], *i.e.*, the fact that a change of one symbol in the plaintext does not reflect into the change of all symbols in the ciphertext. If the encoding process did not entail a dimensionality reduction, this fact could be exploited every time a plaintext-ciphertext pair $\mathbf{x}', \mathbf{y}'$ is known. If a new ciphertext $\mathbf{y}''$ is available and known to lie close to $\mathbf{y}'$, then the corresponding plaintext $\mathbf{x}''$ must be close to $\mathbf{x}'$ thus yielding a good starting point for, *e.g.*, a brute-force *Known-Plaintext Attack (KPA)*.

The fact that $m < n$ would however complicate this setting since the counterimages of $\mathbf{y}''$ through $\mathbf{A}$ belong to a subspace in which points arbitrarily far from $\mathbf{x}'$ exist in principle (*i.e.*, $\mathrm{Ker}\,(\mathbf{A}) \neq \emptyset$). Yet, encoding matrices $\mathbf{A}$ are chosen by design (*i.e.*, by the RIP) so that the probability of their null space aligning with $\mathbf{x}'$ and $\mathbf{x}''$ (that are k-sparse w.r.t. a certain $\mathbf{D}$) is overwhelmingly small.

Hence, even if with a relaxation from the quantitative point of view, neighbouring ciphertexts still strongly hint at neighbouring plaintexts. As an objection to this seemingly unavoidable, issue note that the previous argument only holds when the encoding matrix remains the same for both plaintexts, while by our assumption of Section 6.1.1 on the very large period of the generated sequence of pseudo-random encoding matrices two neighbouring plaintexts $\mathbf{x}', \mathbf{x}''$ will most likely be mapped by different encoding matrices $\mathbf{A}', \mathbf{A}''$ to non-neighbouring ciphertexts $\mathbf{y}', \mathbf{y}''$ by the fact that on-average $mn/2$ symbols will differ between $\mathbf{A}'$ and $\mathbf{A}''$, ensuring a diffusion-like property on the linear encoding performed by CS.

7.2 Achievable Security Properties

The achievable security properties are shown in asymptotic and non-asymptotic configurations of CS, *i.e.*, for $n \to \infty$ and finite n in full analogy with the models in Section 6.2. No guarantee of perfect secrecy[1] is given here, on the basis that caution must be adopted for a linear encoding. We also remark that the presented evidence

---

[1] A recent contribution by Bianchi *et al.* [177] states that perfect secrecy for finite n is achievable by a suitable energy normalisation of the measurements in the RGE encoding matrix case. While this circumvents Proposition 7.1 in the mathematical sense, such a normalisation would imply a loss of relevant information if the energy is not transmitted, otherwise it would delegate the perfect secrecy requirement to a side-channel, leaving its verification completely open.

corresponds to statistical *ciphertext-only attacks* [164] to multiclass encryption by CS.

7.2.1 A Notion and Verification of Asymptotic Secrecy

While perfect secrecy is unachievable, we now introduce the notion of *asymptotic spherical secrecy*[2] and show that CS with i.i.d. RsGE encoding matrices has this property, *i.e.*, no information can be inferred on a plaintext x in model (m_2) from the statistical properties of all its possible ciphertexts but its power. The implication of this property is the basic guarantee that a malicious *eavesdropper* intercepting the measurement vector y will not be able to extract any information on the plaintext except for its power.

Definition 7.1 *(Asymptotic spherical secrecy)*. Let $\mathcal{X}$ be a R.P. whose plaintexts have finite power $0 < W_\mathbf{x} < \infty$, $\mathcal{Y}$ be the R.P. of the corresponding ciphertexts. A cryptosystem has asymptotic spherical secrecy if for any of its plaintexts $\mathbf{x} = \{\mathbf{x}^{(n)}\}_{n=0}^{+\infty}$ and ciphertexts $\mathbf{y} = \{\mathbf{y}^{(m)}\}_{m=0}^{+\infty}$ we have

$$f_{\mathcal{Y}|\mathcal{X}}(\mathbf{y}|\mathbf{x}) \xrightarrow[\text{dist.}]{} f_{\mathcal{Y}|W_\mathbf{x}}(\mathbf{y}) \quad (7.1)$$

where the subscripts of f. indicate the joint and conditional PDFs of the respective R.P.s, $f_{\mathcal{Y}|W_\mathbf{x}}$ denotes conditioning over plaintexts x with identical power $W_\mathbf{x}$, and $\xrightarrow[\text{dist.}]{}$ denotes convergence in distribution as $m, n \to \infty$.

From an eavesdropper's point of view, asymptotic spherical secrecy means that given any ciphertext y we have

$$f_{\mathcal{X}|\mathcal{Y}}(\mathbf{x}|\mathbf{y}) \simeq \frac{f_{\mathcal{Y}|W_\mathbf{x}}(\mathbf{y})}{f_{\mathcal{Y}}(\mathbf{y})} f_{\mathcal{X}}(\mathbf{x})$$

implying that any two different plaintexts with an identical, prior and equal power $W_\mathbf{x}$ will remain approximately indistinguishable from

---

[2]That is a weak form of secrecy, similar in principle to that of Wyner [178], yet posing an emphasis on same-power plaintexts.

ACHIEVABLE SECURITY PROPERTIES

their ciphertexts. In this asymptotic setting, the following proposition holds.

Proposition 7.2 *(Asymptotic spherical secrecy of i.i.d. RsGE encoding matrices).* Let $\mathcal{X}$ be a R.P. with bounded-value plaintexts of finite power $W_\mathbf{x}$, y_j any r.v. in the RP $\mathcal{Y}$ as in (m$_2$). For $n \to \infty$ we have

$$f_{y_j|\mathcal{X}}(y_j) \xrightarrow[\text{dist.}]{} \mathcal{N}(0, W_\mathbf{x}) \qquad (7.2)$$

Thus, i.i.d. RsGE encoding matrices provide independent, asymptotically spherical-secret measurements as in (7.1).

Since the rows of $\mathbf{A}$ are independent, $y_j|W_\mathbf{x}$ are also independent and Proposition 7.2 asserts that, although not secure in the Shannon sense, CS with suitable encoding matrices is able to conceal the plaintext up to the point of guaranteeing its security for $n \to \infty$. The proof of this statement follows.

Proof of Proposition 7.2. The proof is given by simple verification of the Lindeberg-Feller central limit theorem (see [117, Theorem 27.4]) for y_j in $\mathcal{Y}$ conditioned on a plaintext $\mathbf{x}$ of $\mathcal{X}$ in (m$_2$). By the hypotheses, the plaintext $\mathbf{x} = \{x_l\}_{l=0}^{n-1}$ has power $0 < W_\mathbf{x} < \infty$ and $\forall l \in \{0, n-1\}$, $x_l^2 \leq M_\mathbf{x}$ for some finite $M_\mathbf{x} > 0$. Any $y_j|\mathcal{X} = \lim_{n \to \infty} \sum_{l=0}^{n-1} z_{j,l}$ where all $z_{j,l} = A_{j,l} \frac{x_l}{\sqrt{n}}$ is a sequence of independent, non-identically distributed r.v.s of moments $\mu_{z_{j,l}} = 0, \sigma_{z_{j,l}}^2 = \frac{x_l^2}{n}$. By letting the partial sum $S_j^{(n)} = \sum_{l=0}^{n-1} z_{j,l}$, its mean $\mu_{S_j^{(n)}} = 0$ and $\sigma_{S_j^{(n)}}^2 = \frac{1}{n} \sum_{l=0}^{n-1} x_l^2$. Thus, we verify the necessary and sufficient condition [117, (27.19)]

$$\lim_{n \to \infty} \max_{l=0,\ldots,n-1} \frac{\sigma_{z_{j,l}}^2}{\sigma_{S_j^{(n)}}^2} = 0$$

by straightforwardly observing

$$\lim_{n \to \infty} \max_{l=0,\ldots,n-1} \frac{\frac{x_l^2}{n}}{\frac{1}{n} \sum_{l=0}^{n-1} x_l^2} \leq \frac{M_\mathbf{x}}{W_\mathbf{x}} \lim_{n \to \infty} \frac{1}{n} = 0$$

The verification of this condition guarantees that $y_j|\mathcal{X} = \lim_{n\to\infty} S_j^{(n)}$ is normally distributed with $\mu_{y_j} = 0$ and variance

$$\sigma^2_{y_j|\mathcal{X}} = \lim_{n\to\infty} \mathbb{E}[(S_j^{(n)})^2] = W_{\mathbf{x}}$$

yielding (7.2). □

Summarising, the asymptotic regime allowed the derivation of a weak notion of secrecy that shows how the information leakage from the plaintext into the ciphertext is only limited to $W_{\mathbf{x}}$ when the encoding matrices are i.i.d. RsGEs.

7.2.2 A Verification of Non-Asymptotic Secrecy

Since prospective applications of multiclass encryption by CS, and in general CS with i.i.d. RsGE encoding matrices will entail finite-size configurations eventually requiring n on the order of a few hundreds, it is of primary concern to show what security properties are still granted in a non-asymptotic setting. The achievable security properties are tested below by two empirical methods and one theoretical result, that guarantees an extremely sharp rate of convergence to (7.2) for finite n.

Statistical Cryptanalysis by Hypothesis Testing

As a first empirical illustration of the consequences of asymptotic spherical secrecy for finite n, we consider an attack aiming at distinguishing two orthogonal plaintexts $\mathbf{x}'$ and $\mathbf{x}''$ from their encryption (clearly, finite energy must be assumed as in (m$_1$)). The attacker has access to a large number of ciphertexts collected in a set Y' obtained by applying different, randomly generated RBE encoding matrices to a certain $\mathbf{x}'$. Then, the attacker collects another set Y'' of ciphertexts, all of them corresponding either to $\mathbf{x}'$ or to $\mathbf{x}''$, and attempts to distinguish which is the true plaintext between the two.

This reduces the attack to an application of statistical hypothesis testing [130, Section 11.7], the null assumption being that the distribution underlying the statistical samples in Y'' is the same as that

ACHIEVABLE SECURITY PROPERTIES

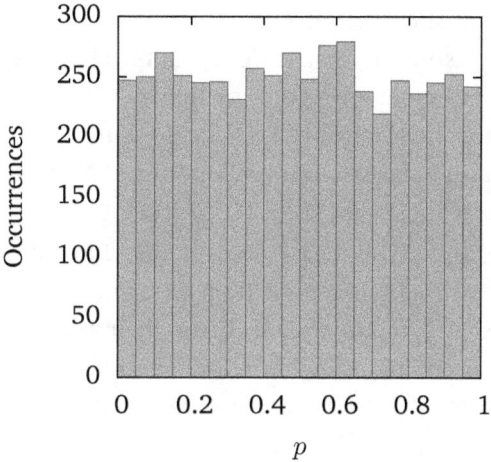

(a) $E_{\mathbf{x}'} = E_{\mathbf{x}''} = 1$; uniformity test value $p = 0.4775$ implies uniformity at 5% significance.

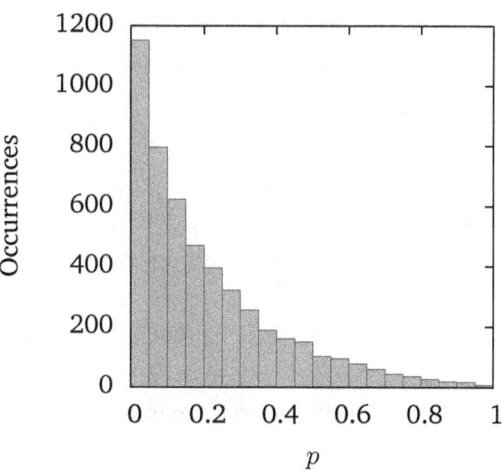

(b) $E_{\mathbf{x}'} = 1$, $E_{\mathbf{x}''} = 1.01$; uniformity test value $p \simeq 0$ implies non-uniformity.

Figure 7.1: Outcome of second-level KS statistical tests to distinguish between two orthogonal plaintexts $\mathbf{x}', \mathbf{x}''$; in (a) $\mathbf{x}', \mathbf{x}''$ have $E_{\mathbf{x}'} = E_{\mathbf{x}''}$, spherical secrecy applies and the uniform distribution of p-values shows that the corresponding ciphertexts are statistically indistinguishable. In (b) $\mathbf{x}', \mathbf{x}''$ have $E_{\mathbf{x}'} \neq E_{\mathbf{x}''}$, spherical secrecy does not apply and the distribution of p-values shows that the corresponding ciphertexts are distinguishable.

underlying the statistical samples in Y'. For maximum reliability we adopt a two-level testing approach: we repeat the above experiment for many instances of random orthogonal plaintexts $\mathbf{x}'$ and $\mathbf{x}''$, performing a two-way *Kolmogorov-Smirnov (KS)* test to compare the empirical distributions obtained from Y' and Y'' produced by such orthogonal plaintexts.

Each of the above KS tests yields a p-value quantifying the probability that two data sets coming from the same distribution exhibit larger differences w.r.t. those at hand. Given their meaning, individual p-values could be compared against a desired significance level to give a first assessment whether the null hypothesis (*i.e.*, equality in distribution) can be rejected.

Yet, since it is known that p-values of independent tests on distributions for which the null assumption is true must be uniformly distributed in $[0,1]$ we collect P of them and feed this second-level set of samples into a one-way KS test to assess uniformity at the standard significance level 5%.

This testing procedure is done for $n = 256$ in the cases $E_{\mathbf{x}'} = E_{\mathbf{x}''} = 1$ (same energy plaintexts) and $E_{\mathbf{x}'} = 1, E_{\mathbf{x}''} = 1.01$, *i.e.*, with a 1% difference in energy between the two plaintexts. The resulting p-values for $P = 5000$ are computed by matching pairs of sets containing $5 \cdot 10^5$ ciphertexts, yielding the p-value histograms depicted in Fig. 7.1. We report these empirical PDFs of the p-values in the two cases along with the p-value of the second-level assessment, *i.e.*, the probability that samples from a uniform distribution exhibit a deviation from a flat histogram larger than the observed one. When the two plaintexts have the same energy, all evidence concurs to say that the ciphertext distributions are statistically indistinguishable. In the second case, even a small difference in energy causes statistically detectable deviations and leads to a correct inference of the true plaintext between the two.

Statistical Cryptanalysis by the Kullback-Leibler Divergence

To reinforce even further the fact that any two plaintexts $\mathbf{x}', \mathbf{x}'' \in \mathbb{R}^n$ under different i.i.d. RsGE encoding matrices cannot be inferred by a

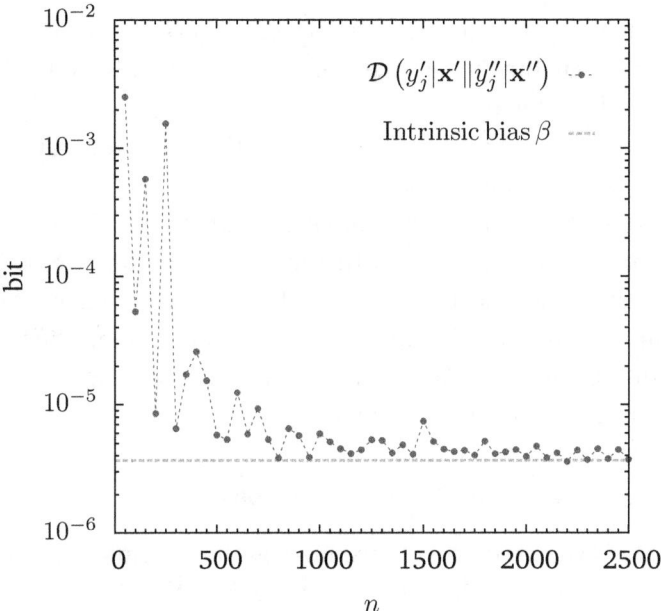

Figure 7.2: Estimated Kullback-Leibler divergence between the probability distributions of two ciphertext elements corresponding to different original signals.

statistical analysis of their ciphertexts $\mathbf{y}', \mathbf{y}' \in \mathbb{R}^m$ even for finite n, we here attempt to do so by recalling the *Kullback-Leibler divergence*[3] [130, (8.46)] of any two r.v.s a, b, i.e.,

$$\mathcal{D}(a\|b) = \int_{-\infty}^{+\infty} f(a) \log \left(\frac{f(a)}{f(b)} \right) \mathrm{d}a \mathrm{d}b \qquad (7.3)$$

that is a simple measure of similarity between the PDF of a and b. We now evaluate $\mathcal{D}(y'_j|\mathbf{x}'\|y''_j|\mathbf{x}'')$ in (7.3) by considering two plaintexts $\mathbf{x}', \mathbf{x}'' \in \mathbb{R}^{2500}$ and extracting sequences of $n = \{50, 100, 150, \ldots, 2500\}$ samples from each of them. For every n the two sample collections are normalised to $E_{\mathbf{x}'} = E_{\mathbf{x}''} = 1$ and projected along 10^8 i.i.d. R.V.s drawn as rows of a RBE, forming a large set of instances of $y'_j|\mathbf{x}', y''_j|\mathbf{x}''$.

---

[3]This quantity is closely related to the concept of mutual information [130, Section 8.5], which is also used in the basic proof of non-perfect secrecy of CS given by Rachlin et al. [26].

These samples are used to form the empirical[4] PDF $\hat{f}(y'_j|\mathbf{x}'), \hat{f}(y''_j|\mathbf{x}'')$ and thus estimate the Kullback-Leibler divergence that is plotted in Fig. 7.2 against the value of n.

As a reference, we also report the theoretical expected value of the divergence estimated using two sets of n samples drawn from $\mathcal{N}(0,1)$, *i.e.*, due to the bias of the histogram estimator $\beta \simeq 3.67 \times 10^{-6}$ bit. It is clear that the distributions of the ciphertexts become statistically indistinguishable for n above a few hundreds, since the number of bits of information that can be apparently inferred from their differences (about 10^{-5} bit for $n > 500$) is mainly due to the bias β and thus cannot support a statistical cryptanalysis.

Rate of Convergence for Finite Dimensions

By now, we have observed with two methods how asymptotic spherical secrecy has finite n effects; from a more formal point of view, we now evaluate the convergence rate of (7.2) for finite n to conclude with a guarantee that an eavesdropper intercepting the ciphertext will observe samples of an approximately Gaussian R.V. bearing very little information in addition to the energy of the plaintext. We hereby consider $\mathbf{x}$ a R.V. as in (m_1), for which a plaintext $\mathbf{x}$ of energy $E_\mathbf{x}$ lies on the sphere $\Sigma_{E_\mathbf{x}}^{n-1} = \{\mathbf{x} \in \mathbb{R}^n : \|\mathbf{x}\|_2^2 \leq E_\mathbf{x}^2\}$. The procedure to verify the rate of convergence of (7.2) in this specific case substantially requires a study of the distribution of a linear combination of r.v.s, $y_j = \sum_{l=0}^{n-1} A_{j,l} x_l$ conditioned on $\mathbf{x} = \{x_l\}_{l=0}^{n-1} \in \Sigma_{E_\mathbf{x}}^{n-1}$.

The most general convergence rate for sums of i.i.d. r.v.s is given by the well-known Berry-Esseen Theorem [179] as $\mathcal{O}(n^{-\frac{1}{2}})$. In our case we apply a recent, remarkable result of [180] that improves and extends this convergence rate, *i.e.*, that addresses the case of inner products of i.i.d. R.V.s (*i.e.*, any row of $\mathbf{A}$) and vectors (*i.e.*, the plaintexts $\mathbf{x}$) uniformly distributed on $\Sigma_{E_\mathbf{x}}^{n-1}$.

---

[4]In order to enhance this evaluation, an optimal non-uniform binning is applied in the estimation of the histograms, since the PDFs are expected to be distributed as $\mathcal{N}(0,1)$. This binning amounts to taking the inverse *Cumulative Distribution Function (CDF)* of the standard normal distribution to obtain 256 uniform-probability bins, thus maximising their entropy.

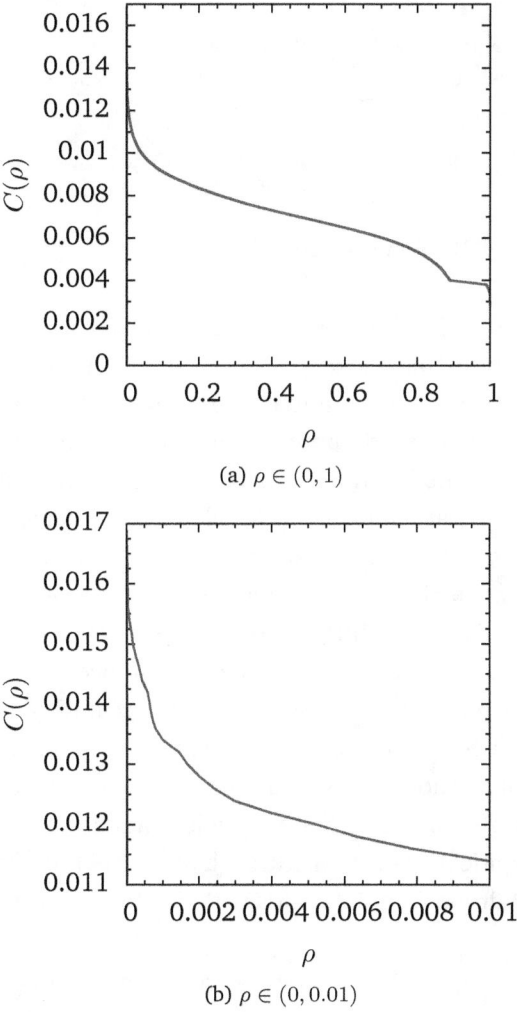

Figure 7.3: Empirical evaluation of $C(\rho)$ in the convergence rate (7.4) based on a large number of plaintexts $\mathbf{x}$ on the sphere Σ_1^{n-1} and $n = 2^4, 2^5, \ldots, 2^{10}$.

Proposition 7.3 *(Rate of convergence with i.i.d. RsGE encoding matrices).* Let $\mathbf{x}, \mathbf{y}$ be R.V.s as in (m_1) with $\mathbf{A}$ drawn from a RsGE with i.i.d. entries having zero-mean, unit-variance, finite fourth-moment entries. For any $\rho \in (0,1)$, there exists a subset $\mathcal{B} \subseteq \Sigma_{E_\mathbf{x}}^{n-1}$ with a probability measure $\sigma^{n-1}(\mathcal{B}) \geq 1 - \rho$ such that all y_j in $\mathbf{y}$ verify

$$\sup_{\alpha < \beta} \left| \int_\alpha^\beta f(y_j | \mathbf{x} \in \mathcal{B}) dy_j - \frac{1}{\sqrt{2\pi}} \int_\alpha^\beta e^{-\frac{t^2}{2E_\mathbf{x}}} dt \right| \leq \frac{C(\rho)}{n} \quad (7.4)$$

for $C(\rho)$ a non-increasing function of ρ.

Proposition 7.3 with ρ sufficiently small means that it is most likely (actually, with probability exceeding $1 - \rho$) to observe an $\mathcal{O}(n^{-1})$ convergence between $f(y_j | \mathbf{x})$ and the limiting distribution $\mathcal{N}(0, E_\mathbf{x})$. The function $C(\rho)$ is loosely bounded in [180], so to complete this analysis we performed a thorough Monte Carlo evaluation of its possible values. In particular, we have taken 10^4 instances of a R.V. $\mathbf{x}$ uniformly distributed on Σ_1^{n-1} for each $n = 2^4, 2^5, \ldots, 2^{10}$. The PDF $f(y_j | \mathbf{x})$ is estimated with the following procedure: we generate $5 \cdot 10^7$ rows of an i.i.d. RBE encoding matrix and perform the usual linear encoding, thus yielding the same number of instances of y_j for each $\mathbf{x}$ and n. On this large sample set we are able to accurately estimate the previous PDF on 4096 equiprobable intervals, and compare it to the same binning of the normal distribution as in the *left-hand side (LHS)* of (7.4) for each (x, n). This method yields sample values for (7.4), allowing an empirical evaluation of the quantity $C(\rho)$. In this example, when $\rho \geq 10^{-3}$ Proposition 7.3 holds with $C(\rho) = 1.34 \cdot 10^{-2}$.

Proof of Proposition 7.3. We start by considering y_j in $\mathbf{y}$ of model (m_1) conditioned on a given $\mathbf{x}$ with finite energy $E_\mathbf{x}$. Each of such variables is a linear combination of n i.i.d. r.v.s $A_{j,l}$ with zero-mean, unit-variance and finite fourth-moments. The coefficients of this linear combination are the plaintext $\mathbf{x}$, which by now we assume to have $E_\mathbf{x} = 1$, *i.e.*, to lie on the unit sphere Σ_1^{n-1} of $\mathbb{R}^n$. Define $\gamma = \left(\frac{1}{n} \sum_{l=0}^{n-1} \mathbb{E}[A_{j,l}^4] \right)^{\frac{1}{4}} < \infty$, which for RBE matrices is $\gamma = 1$, whereas for RGE matrices $\gamma = 3^{\frac{1}{4}}$. This setting verifies [180, Theorem 1.1]: for any $\rho \in (0,1)$ there exists a subset $\mathcal{B} \subseteq \Sigma_1^{n-1}$ with a probability measure $\mu(\mathcal{B})$ such that $\sigma^{n-1}(\mathcal{B}) = \frac{\mu(\mathcal{B})}{\mu(\Sigma_1^{n-1})} \geq 1 - \rho$ and if $\mathbf{x} \in \mathcal{B}$, then

$$\sup_{\substack{(\alpha,\beta) \in \mathbb{R}^2 \\ \alpha < \beta}} \left| \mathbb{P}\left[\alpha \leq \sum_{l=0}^{n-1} A_{j,l} x_l \leq \beta \right] - \frac{1}{\sqrt{2\pi}} \int_\alpha^\beta e^{-\frac{t^2}{2}} \, dt \right| \leq \frac{C(\rho)\gamma^4}{n}$$

(7.5)

with $C(\rho)$ a positive, non-increasing function. An application of this result to $\mathbf{x}$ with energy $E_\mathbf{x}$, i.e., on the sphere of radius $\sqrt{E_\mathbf{x}}$, $\gamma = 1$ (A RGE) can be done by straightforwardly scaling the standard normal PDF in (7.5) to $\mathcal{N}(0, E_\mathbf{x})$, thus yielding the statement of Proposition 7.3. □

7.2.3 Statistical Cryptanalysis and Multiclass Encryption

Contextualising the above findings to multiclass encryption by CS, we have shown how a malicious eavesdropper attempting to break the encoding by means of a straightforward statistical analysis of $\mathbf{y}$ is effectively presented with Gaussian-distributed ciphertexts when the encoding matrix is drawn from an i.i.d. RsGE.

In addition, one could consider the threat of a malicious second-class user attempting to upgrade itself to the knowledge of the true encoding matrix $\mathbf{A}^{(1)}$ given $\mathbf{A}^{(0)}$. Letting $\mathbf{A}^{(0)}, \mathbf{A}^{(1)}$ be drawn from RBEs, in the worst-case we may also assume that this attacker has access to $\epsilon = \Delta \mathbf{A} \mathbf{x}$, and is able to compute $f(\epsilon)$ for a statistical cryptanalysis. Clearly, this will depend on the density of $\Delta \mathbf{A} = \mathbf{A}^{(1)} - \mathbf{A}^{(0)}$, that is a SSF drawn from a RME with i.i.d. entries. Informally and intuitively, this will result in $f(\epsilon|\mathbf{x}) \longrightarrow_{\text{dist.}} \mathcal{N}(\mathbf{0}_m, \mathbf{K}_\epsilon)$ where $\mathbf{K}_\epsilon = \sigma_{\Delta \mathbf{A}}^2 E_\mathbf{x} \mathbf{I}_m$ where $\sigma_{\Delta \mathbf{A}}^2 = 4\eta$ and $E_\mathbf{x} = \|\mathbf{x}\|_2^2$, i.e., the information that leaks to a malicious second-class user is the SSF density η as well as the energy of the plaintext. A more thorough verification can be derived by application of the procedures detailed in this Section. Hence, the ciphertext is statistically indistinguishable from the one that could be produced by encoding the same plaintext with $\mathbf{A}^{(0)}$ instead of $\mathbf{A}^{(1)}$, and such second-class users will be unable to exploit the statistical properties of **y** to upgrade their encoding matrix to $\mathbf{A}^{(1)}$.

Thus, we may safely conclude that straightforward statistical attacks to multiclass encryption based on CS only extract very limited information from the ciphertext; the more threatening case of Known-Plaintext Attacks is expanded in the next Chapter.

Summary

> ▶ CS is not perfectly secret in the Shannon sense in general. However, a definition of asymptotic secrecy allows one to specify which feature of the plaintext leaks into the ciphertext, *i.e.*, the power of the plaintext as $n \to \infty$. Thus, we have shown how CS solely leaks this information in the asymptotic case.

> ▶ In the more concerning non-asymptotic case in which practical encryption based on CS will operate, we have shown that different approaches to the statistical analysis of the ciphertexts yield no information but the energy of the plaintext.

> ▶ An $\mathcal{O}(n^{-1})$ convergence rate to the limiting distribution of the measurements, *i.e.*, a simple Gaussian distribution was shown to hold for any i.i.d. RsGE encoding matrix, thus implying that an eavesdropper performing a statistical cryptanalysis is presented with i.i.d. Gaussian ciphertexts whose variance depends on the energy of the plaintext. Since this information is not sufficient for performing a cryptanalysis, we may consider multiclass encryption by CS a reliable method for non-critical security applications.

A Computational Cryptanalysis of Compressed Sensing

DESPITE the linearity of its encoding, we have shown how CS provides a limited form of secrecy when i.i.d. RsGE encoding matrices are used to produce sets of ciphertexts (*i.e.*, measurements). In this Chapter we quantify the resistance of the least complex form of this kind of encoding, *i.e.*, CS with RBE encoding matrices, against *Known-Plaintext Attacks (KPAs)*. These represent the most threatening form of cryptanalysis such a scheme will suffer. The properties and results of these attacks are fully explored here by theoretical means, as they can be mapped to a combinatorial optimisation problem that models the most informed attack a malicious user may attempt.

For both standard CS and its multiclass encryption embodiment, we show how the average number of candidate encoding matrix rows that match a plaintext-ciphertext pair is huge, thus making the search for the true encoding matrix inconclusive. Such a conclusion was anticipated by [26, 163], where the presented evidence essentially addressed brute-force enumeration; the main difference with our approach is that our quantification is theoretical, and yet matches with surprising precision the odds of empirical attacks. Thus, the findings support a notion of computational security for CS-based encryption schemes.

Still in computational security terms, since missing information on the encoding matrices might be treated as a perturbation matrix, we attempt an additional computational attack specifically targeted to a multiclass scheme and attempting to nullify its effect. This form of attack is carried out by a second-class user that attempts to upgrade its knowledge by using signal recovery algorithms specifically accounting for encoding matrix uncertainty [159, 160]. As expected from the random nature of the perturbation introduced in Chapter 6 the results will however show no practical improvement w.r.t. the bounds and performances illustrated in Section 6.3.

Practical computational attacks are then exemplified by applying CS as an encryption scheme to the same signal classes of Chapter 6, showing how the extracted information on the true encoding matrix from a plaintext-ciphertext pair leads to no significant signal recovery quality increase. This theoretical and empirical evidence clarifies that, although not perfectly secure, both standard CS and multiclass encryption based on it feature a noteworthy level of security against KPAs, thus increasing its appeal as a zero-cost encryption method for resource-limited sensor nodes.

8.1 A Theory for Known-Plaintext Attacks

We here focus on encoding matrices $\mathbf{A}^{(0)}, \mathbf{A}^{(1)} \in \{-1, +1\}^{m \times n}$ drawn from the RBE, as they are remarkably simple and therefore suitable to be generated, implemented and stored in digital devices. Due to their simplicity these matrices are more easily subject to cryptanalysis; on the contrary, if many symbols were used in each element of $\mathbf{A}^{(0)}, \mathbf{A}^{(1)}$ this would cause a rapid consumption of the bits generated by expansion of the secret as discussed in Section 6.1.1. Thus, the RBE encoding matrix case and its use in two-class encryption by CS serves as a basic reference for other RMEs and more complex configurations of multiclass encryption by CS.

To understand the relevance of the security issues addressed in this Section, let us consider a first sequence of matrices $\{(\mathbf{A}^{(0)})_t\}_{t \in \mathbb{Z}}$ obtained by pseudo-random expansion of a seed $\text{Key}(\mathbf{A}^{(0)})$. In parallel, a sequence of index pair sets $\{(C^{(0)})_t\}_{t \in \mathbb{Z}}, (C^{(0)})_t \subseteq$

$\{0,\ldots,m-1\}\times\{0,\ldots,n-1\}$ is obtained by pseudo-random expansion of a seed Key $(C^{(0)})$. We then generate a second sequence of matrices $\{(\mathbf{A}^{(1)})_t\}_{t\in\mathbb{Z}}$ whose elements $(\mathbf{A}^{(1)})_t$ are obtained by combining $(\mathbf{A}^{(1)})_t, (C^{(0)})_t$ by (6.3). Clearly, the strong assumption that any encoding matrix is never reused in the encoding (Section 6.1.1) is incompatible with the use of such sequences, as they will eventually repeat due to their pseudo-random nature; nevertheless, we may assume that the sequences' period is sufficiently long to avoid repetition in the attacker's observation time. But even with this assumption standing, if an attacker was able to recover even a few elements in the above matrix sequences, this would potentially enable, *e.g.*, PRNG cryptanalysis strategies (*e.g.*, [166] for LFSRs) to break the cipher by retrieving the seeds in Fig. 6.1.

Hence, to avoid such an event we focus on showing that a single, generic instance of $\mathbf{A}^{(1)}$ in $\mathbf{y} = \mathbf{A}^{(1)}\mathbf{x}$ cannot be recovered even with the highest level of information, *i.e.*, given $\mathbf{x}$ and $\mathbf{y}$. We consider for the sake of simplicity a fixed cardinality c for every $(C^{(0)})_t$, recall that $\eta = c/mn$ is the SSF density (*i.e.*, the number of non-zeros in $\mathbf{\Delta A} = \mathbf{A}^{(1)} - \mathbf{A}^{(0)}$), and let $\mathbf{A}^{(0)}, \mathbf{A}^{(1)}, C^{(0)}$ be generic, unique instances of the above sequences of pseudo-random matrices and index pair sets respectively.

Thus, we are considering a threatening situation in which an attacker has gained access to a known plaintext $\mathbf{x}$ corresponding to a known ciphertext $\mathbf{y}$. Based on these priors, the attacker aims at computing the true encoding $\mathbf{A}^{(1)}$ by carrying out a KPA. In the following we will consider this attack by assuming that *only one* $(\mathbf{x}, \mathbf{y})$ pair is known for a certain $\mathbf{A}^{(1)}$, consistently with the hypothesis that the same $\mathbf{A}^{(1)}$ only reappears after a long period[1].

Starting from a single pair $(\mathbf{x}, \mathbf{y})$, depending on the level of information available to the attacker we obtain two KPA perspectives (see Fig. 8.1): the first is that of a pure eavesdropper, Eve, and addresses the problem of retrieving $\mathbf{A}^{(1)}$ given $(\mathbf{x}, \mathbf{y})$; the second is

---

[1]Note that if n independent $(\mathbf{x}, \mathbf{y})$ pairs were known for the same $\mathbf{A}^{(1)}$, one could resort to elementary linear algebra and infer the true encoding matrix by solving a simple linear system.

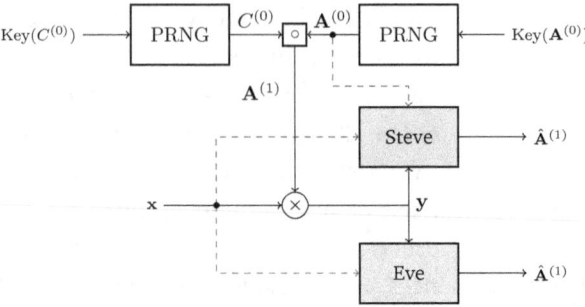

Figure 8.1: A two-class encryption scheme and the known-plaintext attacks being analysed from an eavesdropper (Eve) and a second-class user (Steve).

that of a second-class receiver, Steve, that faces the same problem of retrieving $\mathbf{A}^{(1)}$ given $(\mathbf{x}, \mathbf{y}, \mathbf{A}^{(0)})$, where $\mathbf{A}^{(0)}$ is partially correct and chosen as in the two-class scheme of Chapter 6. This second form is also referred to as *class-upgrade*.

Since the attacks we discuss rely on deterministic knowledge of $\mathbf{x}$ and $\mathbf{y}$, we assume throughout this Chapter that both plaintexts and ciphertexts are represented by digital words. This quantisation is unavoidable as $\mathbf{x}$ and $\mathbf{y}$ will be stored and processed on a digital architecture from which the attacks are carried out. For simplicity, we let $\mathbf{x} = \{x_l\}_{l=0}^{n-1}$ be such that $x_l \in \{-L, \ldots, -1, 0, 1, \ldots, L\}$ for some $L \in \mathbb{Z}_+$ (i.e., we take $\mathbf{x} = \tilde{\mathbf{x}}$ in the fashion of Chapter 3). Note that the number of bits representing the plaintext in this fashion is at least $b_{\mathbf{x}} = \lceil \log_2(2L+1) \rceil$, so we may assume $b_{\mathbf{x}}$ is less than a few tens in typical embodiments (actually, $b_{\mathbf{x}} \leq 16\,\text{bit}$ if the plaintext was previously generated by a common A/D converter). Consequently, the ciphertext will be represented by $\mathbf{y} = \{y_l\}_{l=0}^{m-1}$, each y_l quantised with as many bits as needed to avoid any information loss. In this necessarily digital-to-digital perspective, we will see how the solutions in $\mathbf{A}^{(1)}$ are also a function of the number of bits representing the plaintext (and consequently the ciphertext).

Our KPA analysis applies on a single row[2] of $\mathbf{A}^{(1)}$, i.e., $\mathbf{A}^{(1)}_j$.

---

[2] $\mathbf{A}_j$ here denotes the j-th row of a matrix $\mathbf{A}$.

Furthermore, we note that the analysis is carried out in full compliance with Kerckhoffs's principle [181], *i.e.*, , the only information that the attackers are missing is their respective part of the encryption key, while any other detail on the sparsity basis, as well as two-class encryption specifications is here regarded as known. The actual breaking of the encryption protocol would entail iterating the following attack for all m rows of many of the matrices in the sequence, thus requiring an even larger effort than the one described below. Nevertheless, even knowing one row without uncertainty could lead to a decryption of the pseudo-random sequence generating $\mathbf{A}^{(1)}$, hence the relevance of this simplified case.

8.1.1 Eavesdropper's Known-Plaintext Attack

Given a plaintext $\mathbf{x}$ and the corresponding ciphertext $\mathbf{y} = \mathbf{A}^{(1)}\mathbf{x}$ we now assume the perspective of Eve and attempt to recover $\mathbf{A}_j^{(1)}$ with a set of symbols $\hat{\mathbf{A}}_j^{(1)} = \{\hat{A}_{j,l}^{(1)}\}_{l=0}^{n-1} \in \{-1,+1\}^n$ such that the j-th symbol in the ciphertext,

$$y_j = \sum_{l=0}^{n-1} A_{j,l}^{(1)} x_l = \sum_{l=0}^{n-1} \hat{A}_{j,l}^{(1)} x_l \tag{8.1}$$

Moreover, to favour the attacker[3] we assume all $x_l \neq 0$. We now introduce a combinatorial optimisation problem at the heart of the analysed KPAs.

Problem 8.1 (*Subset-Sum Problem*). Let $\{u_l\}_{l=0}^{n-1}, u_l \in \{1,\ldots,L\}$ and $v \in \mathbb{Z}_+$. We define *Subset-Sum Problem (SSP)* [182, Chap. 4] the problem of assigning n binary variables $b_l \in \{0,1\}, l = \{0,\ldots,n-1\}$ so that

$$v = \sum_{l=0}^{n-1} b_l u_l \tag{8.2}$$

---

[3] If any $x_l = 0$ each corresponding term would give no contribution to the sum (8.1), thus making $\hat{A}_{j,l}^{(1)}$ an undetermined variable in the attack. Hence, the sparsity of $\mathbf{x}$ would actually be an issue for the attacker, which is why the sparsity basis $\mathbf{D}$ never appears in the present evaluation.

We define *solution* any $\{b_l\}_{l=0}^{n-1}$ verifying (8.2). In this configuration, the *density* of this combinatorial problem is defined as [183]

$$\delta(n, L) = \frac{n}{\log_2 L} \qquad (8.3)$$

Although in general a SSP is NP-complete, not all of its instances are equally hard. In fact, it is known that *high-density* instances (*i.e.*, with $\delta(n, L) > 1$) have plenty of solutions found or approximated by, e.g., dynamic programming, whereas *low-density* instances are harder, although for special cases polynomial-time algorithms have also been found [183]. On a historical note, such low-density hard SSP instances have been used in cryptography to develop the family of *public-key knapsack cryptosystems* [184, 185] although most have been broken with polynomial-time algorithms [186]. Problem 8.1 finds a direct application to model Eve's KPA as follows.

Proposition 8.1 (*Eve's Known-Plaintext Attack*). The KPA to $\mathbf{A}_j^{(1)}$ given $(\mathbf{x}, \mathbf{y})$ is equivalent to a SSP where each $u_l = |x_l|$, the variables

$$b_l = \frac{1}{2}\left(\text{sign}(x_l)\,\hat{A}_{j,l}^{(1)} + 1\right)$$

and the sum

$$v = \frac{1}{2}\left(y_j + \sum_{l=0}^{n-1}|x_l|\right)$$

This SSP has a *true solution* $\{\bar{b}_l\}_{l=0}^{n-1}$ that is mapped to the row $\mathbf{A}_j^{(1)}$, and other *candidate solutions* that verify (8.2) but correspond to matrix rows $\hat{\mathbf{A}}_j^{(1)} \neq \mathbf{A}_j^{(1)}$.

We also define $(\mathbf{x}, \mathbf{y}, \mathbf{A}_j^{(1)})$ a *problem instance*. This mapping is obtained as follows.

Proof of Proposition 1. Define the binary variables $b_l \in \{0, 1\}$ so that $\text{sign}(x_l)\,\hat{A}_{j,l}^{(1)} = 2b_l - 1$ and the positive coefficients $u_l = |x_l|$. With this choice (8.1) is equivalent to $y_j = \sum_{l=0}^{n-1}(2b_l - 1)u_l$ which leads to a SSP with $v = \frac{1}{2}\left(y_j + \sum_{l=0}^{n-1}|x_l|\right)$. Since we know that each ciphertext entry y_j must correspond to the inner product between $\mathbf{x}$ and the row $\mathbf{A}_j^{(1)}$, the latter's entries are straightforwardly mapped to the *true* solution of this SSP, i.e., $\{\bar{b}_l\}_{l=0}^{n-1}$. □

In our case we see that the density (8.3) is high since n is large and $\log_2 L$ is fixed by the digital representation of $\mathbf{x}$ (e.g., so that $b_\mathbf{x} \leq 64$). We are therefore operating in a high-density region of problem (8.2). In fact, the resistance of the analysed embodiment of CS to KPAs is not due to the hardness of the corresponding SSP but, as we show below, to the huge number of candidate solutions as n increases, among which an attacker should find the true solution to guess a single row of $\mathbf{A}^{(1)}$. Since no *a priori* criterion exists to select them, we consider them *indistinguishable*.

The next Theorem calculates the expected number of candidate solutions to Eve's KPA by applying the theory developed in [187].

Theorem 8.1 *(Expected number of solutions for Eve's KPA)*. For large n, the expected number of candidate solutions of the KPA in Proposition 8.1, in which (i) all the coefficients $\{u_l\}_{l=0}^{n-1}$ are i.i.d. uniformly drawn from $\{1,\ldots,L\}$, and (ii) the true solution $\{\bar{b}_l\}_{l=0}^{n-1}$ is drawn with equiprobable and independent binary values, is

$$\mathcal{S}_{\text{Eve}}(n, L) \simeq \frac{2^n}{L}\sqrt{\frac{3}{\pi n}} \qquad (8.4)$$

The proof of Theorem 8.1 is given in the next Section. This result (as well as the whole statistical mechanics framework from which it is derived) gives no hint on how much (8.4) is representative of finite-n behaviours. To compensate for that, we enumerated the solutions of several randomly generated small-n problem instances by using CPLEX as a binary programming solver [94] and forcing the computation of the full solution pool; this allowed a verification of the asymptotic expression of (8.4) by comparing its expected number of solutions with those effectively yielded by a computational implementation of Eve's KPA.

Such numerical evidence is reported in Fig. 8.2, where the empirical average number of solutions $\hat{\mathcal{S}}_{\text{Eve}}(n, L)$ to 50 problem

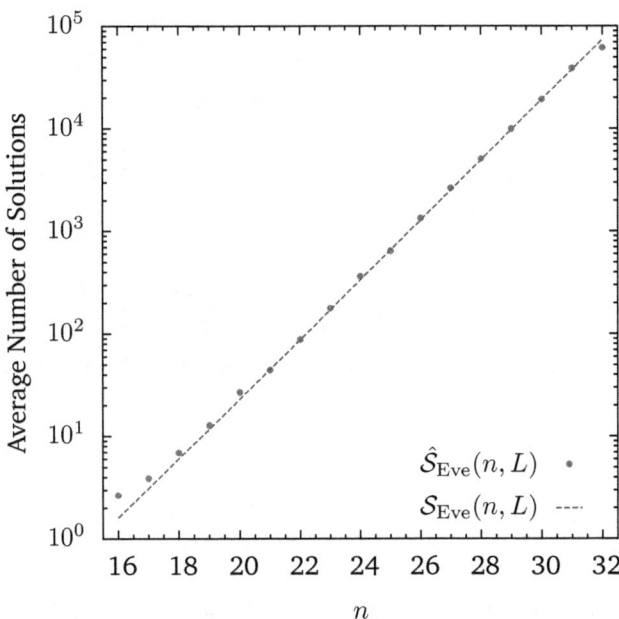

Figure 8.2: Empirical average number of solutions for Eve's KPA compared to the theoretical approximation of (8.4) for $L = 10^4$.

instances with $L = 10^4$ and $n = \{16, \ldots, 32\}$ is plotted and compared with (8.4). The remarkable matching observed there allows us to estimate, for example, that a KPA to the encoding of a grayscale image of $n = 64 \times 64$ pixel quantised with $b_\mathbf{x} = 8$ bit (unsigned) would have to discriminate on the average between $1.25 \cdot 10^{1229}$ equally good candidate solutions for each of the rows of the encoding matrix. This number is not far from the total possible rows, $2^{4096} = 1.04 \cdot 10^{1233}$. Hence, any attacker using this strategy is faced with a deluge of candidate solutions, from which it would choose one presumed to be a piece of the encoding matrix to attempt a guess on $\mathbf{A}^{(1)}$.

A Proof of Theorem 8.1

Firstly, we introduce a technical definition that is used in the proof of Theorem 8.1, as well as that of Theorem 8.3.

Definition 8.1. We define the functions

$$F_p(a,b) = \int_0^1 \frac{\xi^p}{1+e^{a\xi-b}}\,\mathrm{d}\xi \tag{8.5}$$

$$G_p(a,b) = \int_0^1 \frac{\xi^p}{\left(1+e^{a\xi-b}\right)\left(1+e^{b-a\xi}\right)}\,\mathrm{d}\xi \tag{8.6}$$

We now proceed to proving the main statement by means of an interface with the theory developed by Sasamoto et al. [187] on the number of solutions of the SSP.

Proof of Theorem 8.1. Let us first note that, for large n, v in Proposition 8.1 is an integer in the range $[0, nL/2]$, with the values outside this interval being asymptotically unachievable as $n \to \infty$ (see [187, Section 4]). We let $\tau = v/nL$, $\tau \in [0, 1/2]$, and $a(\tau)$ be the solution in a of the equation $\tau = F_1(a,0)$ (i.e., [187, (4.2)]) that is unique since $F_p(a,0)$ in (8.5) is monotonically decreasing in a.

From [187, (4.1)] the number of solutions of a SSP with integer coefficients $\{u_l\}_{l=0}^{n-1}$ uniformly distributed in $[1, L]$ is

$$\mathcal{S}_{\text{Eve}}(\tau, n, L) \simeq \frac{e^{n\left[a(\tau)\tau + \int_0^1 \log\left(1+e^{-a(\tau)\xi}\right)\mathrm{d}\xi\right]}}{\sqrt{2\pi n L^2 G_2(a(\tau), 0)}}$$

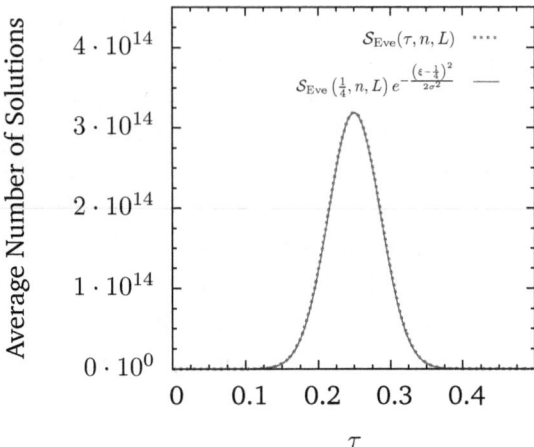

Figure 8.3: Gaussian approximation of $\mathcal{S}_{\text{Eve}}(\tau, n, L)$ for $n = 64$, $L = 10^4$ by letting $\sigma^2 \approx 1/12n$.

that we anticipate to have an approximately Gaussian profile (see Fig. 8.3).

We now compute the average of $\mathcal{S}_{\text{Eve}}(\tau, n, L)$ in τ, that clearly depends on the probability of selecting any value of $v \in [0, \frac{nL}{2}]$, i.e., of $\tau \in [0, \frac{1}{2}]$. Since v is the result of a linear combination, the probability that a specific value appears in a random instance of the SSP is proportional to the number of solutions associated to it. In normalised terms, the PDF of τ must be proportional to $\mathcal{S}_{\text{Eve}}(\tau, n, L)$, i.e., τ is distributed as

$$f_\tau(t) = \frac{1}{\int_0^{\frac{1}{2}} \mathcal{S}_{\text{Eve}}(\xi, n, L) d\xi} \begin{cases} \mathcal{S}_{\text{Eve}}(t, n, L), & 0 \leq t \leq \frac{1}{2} \\ 0, & \text{otherwise} \end{cases}$$

With $f_\tau(t)$ we can compute the expected number of solutions:

$$\mathbb{E}_\tau[\mathcal{S}_{\text{Eve}}(\tau, n, L)] = \frac{\int_0^{\frac{1}{2}} \mathcal{S}_{\text{Eve}}^2(\xi, n, L) d\xi}{\int_0^{\frac{1}{2}} \mathcal{S}_{\text{Eve}}(\xi, n, L) d\xi} \tag{8.7}$$

Although we could resort to numerical integration, (8.7) can be simplified by exploiting what noted above, i.e., that $\mathcal{S}_{\text{Eve}}(\tau, n, L)$ has an approximately Gaussian profile in τ (Fig. 8.3) with a maximum in $\tau = 1/4$. Hence, the expectation in τ becomes

$$\mathbb{E}_\tau[\mathcal{S}_{\text{Eve}}(\tau, n, L)] \simeq \mathcal{S}_{\text{Eve}}\left(\frac{1}{4}, n, L\right) \frac{\int_{-\infty}^{\infty} \left(e^{-\frac{(\xi-\frac{1}{4})^2}{2\sigma^2}}\right)^2 d\xi}{\int_{-\infty}^{\infty} e^{-\frac{(\xi-\frac{1}{4})^2}{2\sigma^2}} d\xi}$$

$$= \mathcal{S}_{\text{Eve}}\left(\frac{1}{4}, n, L\right) \frac{1}{\sqrt{2}} = \frac{2^n}{L}\sqrt{\frac{3}{\pi n}} \qquad (8.8)$$

that is actually independent of the σ^2 used in the Gaussian approximation, and in which we have exploited $a(1/4) = 0$ to obtain the statement of the Theorem. □

8.1.2 Expected Distance of an Eavesdropper's KPA Solutions

A legitimate concern when Eve is presented with a large set of solutions output from a complete KPA to a row of $\mathbf{A}^{(1)}$ is that most of them could be good approximations of the true encoding matrix row. To see whether this is the case, we quantify the difference between $\mathbf{A}_j^{(1)}$ and the corresponding candidate $\hat{\mathbf{A}}_j^{(1)}$ resulting from a KPA in terms of their Hamming distance, *i.e.*, as the number of entries in which they differ.

Theorem 8.2 *(Expected number of solutions for Eve's KPA at a given Hamming distance from the true one).* The expected number of candidate solutions at Hamming distance h from the true solution of the KPA in Proposition 8.1, in which (i) all the coefficients $\{u_l\}_{l=0}^{n-1}$ are i.i.d. uniformly drawn from $\{1, \ldots, L\}$, (ii) the true solution $\{\bar{b}_l\}_{l=0}^{n-1}$ is drawn with equiprobable and independent binary values, is

$$\mathcal{S}_{\text{Eve}}^{(h)}(n, L) = \binom{n}{h} \frac{P_h(L)}{2^h L^h} \qquad (8.9)$$

where $P_h(L)$ is a polynomial in L whose coefficients are reported in Table 8.1 for $h = \{2, \ldots, 15\}$.

The proof of this Theorem and the derivation of Table 8.1 are reported in the next Section. We now collect empirical evidence that the expression in (8.9) correctly anticipates the expected number of solutions at a given Hamming distance. The procedure simply entails processing the enumerated solutions in Section 8.1.1. Thus, Fig. 8.4 reports for $n = \{21, 23, \ldots, 31\}$ the empirical average, over 50 problem instances, of the number of solutions to Eve's KPA whose Hamming distance from the true one is a given value $h = \{2, \ldots, 15\}$, as compared against the value predicted by (8.9) with the polynomial coefficients in Table 8.1. The remarkable matching we observe allows us to estimate that, in the case of a grayscale image ($n = 4096$, $L = 128$), only $1.95 \cdot 10^{41}$ candidate solutions out of the average $1.25 \cdot 10^{1229}$ are expected to have a Hamming distance $h \leq 16$, while $6.33 \cdot 10^{76}$ attain a Hamming distance $h \leq 32$. Since these results apply to each row of the matrix being inferred, this indicates how the chance that a randomly chosen candidate solution is (or is close to) the true one is negligible.

A Proof of Theorem 8.2

We present a proof of the result in (8.9) based on a counting argument.

Proof of Theorem 8.2. We here concentrate on counting the number of candidate solutions $\{b_l\}_{l=0}^{n-1}$ to Eve's KPA that differ from the true one, $\{\bar{b}_l\}_{l=0}^{n-1}$, by exactly h components (at Hamming distance h). We assume that $K \subseteq \{0, \ldots, n-1\}$ is the set of indices for which there is a disagreement, *i.e.*, for all $l \in K$ we have $b_l = 1 - \bar{b}_l$; this set has cardinality h, and is one among $\binom{n}{h}$ possible sets. Since both $\{b_l\}_{l=0}^{n-1}$ and $\{\bar{b}_l\}_{l=0}^{n-1}$ are solutions to the same SSP, and that $b_l = \bar{b}_l$ are identical for $l \notin K$, $\sum_{l \in K} (1 - \bar{b}_l) u_l = \sum_{l \in K} \bar{b}_l u_l$ must hold, implying the equality

$$\sum_{\substack{l \in K \\ \bar{b}_l = 0}} u_l - \sum_{\substack{l \in K \\ \bar{b}_l = 1}} u_l = 0 \qquad (8.10)$$

Although (8.10) recalls the well-known partition problem, in our case K is chosen by each problem instance that sets all u_l and $\bar{b}_l$. Thus,

h	p_1^h	p_2^h	p_3^h	p_4^h	p_5^h	p_6^h	p_7^h	p_8^h	p_9^h	p_{10}^h	p_{11}^h	p_{12}^h	p_{13}^h	p_{14}^h
2	2													
3	-3	3												
4	$\frac{14}{3}$	-4	$\frac{16}{3}$											
5	$-\frac{15}{2}$	$\frac{65}{12}$	$\frac{15}{2}$	$\frac{115}{12}$										
6	$\frac{62}{5}$	$-\frac{15}{2}$	11	$-\frac{27}{2}$	$\frac{88}{5}$									
7	-21	$\frac{959}{90}$	$\frac{203}{12}$	$\frac{707}{36}$	$-\frac{301}{12}$	$\frac{5887}{180}$								
8	$\frac{254}{7}$	$-\frac{140}{9}$	$\frac{1226}{45}$	$-\frac{266}{9}$	$\frac{334}{9}$	$-\frac{422}{9}$	$\frac{19328}{315}$							
9	$-\frac{255}{4}$	$\frac{2613}{112}$	$\frac{731}{16}$	$\frac{14701}{320}$	$-\frac{457}{8}$	$\frac{2233}{32}$	$-\frac{1415}{16}$	$\frac{259723}{2240}$						
10	$\frac{1022}{9}$	$-\frac{2585}{72}$	$\frac{359105}{4536}$	$-\frac{7055}{96}$	$\frac{9869}{108}$	$-\frac{1725}{16}$	$\frac{28625}{216}$	$-\frac{48325}{288}$	$\frac{124952}{567}$					
11	$-\frac{1023}{5}$	$\frac{16973}{300}$	$\frac{60775}{432}$	$\frac{5463953}{45360}$	$-\frac{435941}{2880}$	$\frac{7449761}{43200}$	$-\frac{19811}{96}$	$\frac{1091629}{4320}$	$-\frac{2764663}{8640}$	$\frac{3811773117}{907200}$				
12	$\frac{4094}{11}$	$-\frac{2277}{25}$	$\frac{687791}{2700}$	$-\frac{72523}{360}$	$\frac{3907067}{15120}$	$-\frac{341143}{1200}$	$\frac{599327}{1800}$	$-\frac{7909}{20}$	$\frac{1045349}{2160}$	$-\frac{2205833}{3600}$	$\frac{41931328}{51975}$			
13	$-\frac{1365}{2}$	$\frac{591721}{3960}$	$\frac{2020421}{4320}$	$\frac{44385419}{129600}$	$-\frac{7815847}{17280}$	$\frac{116257063}{241920}$	$-\frac{3192163}{5760}$	$\frac{110721221}{172800}$	$-\frac{13148473}{17280}$	$\frac{19285357}{20736}$	$-\frac{20345507}{17280}$	$\frac{20646903199}{13305600}$		
14	$\frac{16382}{13}$	$-\frac{44863}{180}$	$\frac{343353347}{39600}$	$-\frac{38237381}{64800}$	$\frac{1292711}{1600}$	$-\frac{42972293}{51840}$	$\frac{122732801}{129600}$	$-\frac{92420419}{86400}$	$\frac{53508931}{43200}$	$-\frac{76095383}{51840}$	$\frac{77441609}{43200}$	$-\frac{588168119}{259200}$	$\frac{8667732192}{289575}$	
15	$-\frac{16383}{7}$	$\frac{1074679}{2548}$	$\frac{583763}{360}$	$\frac{1139382839}{110880}$	$-\frac{12673507}{8640}$	$\frac{58584511}{40320}$	$-\frac{400088153}{241920}$	$\frac{1033251187}{564480}$	$-\frac{239927713}{115200}$	$\frac{193398181}{80640}$	$-\frac{98109773}{34560}$	$\frac{273340567}{80640}$	$-\frac{1060663411}{241920}$	$\frac{4671683100097}{80720640}$

Table 8.1: Table of coefficients of the polynomials $P_h(L) = \sum_{j=1}^{h-1} p_j^h L^j$ describing the expected Hamming distance of the solutions to Eve's KPA in (8.9) for $h = \{2, \ldots, 15\}$.

(8.10) holds in a number of cases that depends on how many of the $2^h L^h$ possible assignments of all u_l and $\bar{b}_l$ satisfy it. The only feasible cases are for $h > 1$, and to analyse them we assume $K = \{0, \ldots, h-1\}$ (the disagreements occur in the first h ordered indices) without loss of generality.

Moreover, when (8.10) holds for some $\{\bar{b}_l\}_{l=0}^{n-1}$ it also holds for $\{1-\bar{b}_l\}_{l=0}^{n-1}$. Hence, we may count the configurations that verify (8.10) with $\bar{b}_0 = 0$, knowing that their number will be only *half* of the total. With this, the configurations with $\bar{b}_0 = 0$ must have $\bar{b}_l = 1$ for *at least* one $l > 0$ in order to satisfy (8.10), giving $2^{h-1} - 1$ total cases to check.

The following paragraphs illustrate that, for $h < L$, the number of configurations that verify (8.10) can be written as a polynomial of order $h-1$. With this in mind we can start with the explicit computation for $h = \{2, 3\}$.

1. for $h = 2$, there is only one feasible assignment for the $\{\bar{b}_l\}_{l=0}^{n-1}$, so $u_0 = u_1$ in (8.10), which makes $2L$ cases out of $2^2 L^2$;

2. for $h = 3$, one has 3 feasible assignments for the $\{\bar{b}_l\}_{l=0}^{n-1}$. Due to the symmetry of (8.10) all the configurations have the same behaviour and we may focus on, e.g., $\bar{b}_0 = \bar{b}_1 = 0$ and $\bar{b}_2 = 1 \Rightarrow u_0 + u_1 = u_2$; this can be satisfied only when $u_0 + u_1 \leq L$, i.e., for $L(L-1)/2$ configurations. This makes a total of $2 \cdot 3 \cdot \frac{L(L-1)}{2} = 3L(L-1)$ over the $2^3 L^3$ possible configurations;

3. for $h > 3$, this procedure is much less intuitive; nevertheless, we can at least prove that the function $P_h(L)$ counting the configurations for which (8.10) holds is a polynomial in L of degree $h - 1$. To show this, let us proceed in three steps.

    a) Indicate with $\pi_{\bar{b}}$ the $(h-1)$-dimensional subspace of $\mathbb{R}^h$ defined by $\sum_{\substack{l \in K \\ \bar{b}_l = 0}} \xi_l - \sum_{\substack{l \in K \\ \bar{b}_l = 1}} \xi_l = 0, \xi \in \mathbb{R}^h$. The intersection $\alpha_{\bar{b}}(L) = [1, L]^h \cap \pi_{\bar{b}}$ is such that each assignment of $\{u_l\}_{l=0}^{h-1} \in [1, L]^h$ satisfying (8.10) is an integer point in $\alpha_{\bar{b}}$. To count those points define $\beta_{\bar{b}}(L) = [0, L+1] \cap \pi_{\bar{b}}$ and note that the number of integer points in $\alpha_{\bar{b}}$ is equal to the number of integer points in the interior of $\beta_{\bar{b}}$ (the points on the frontier of $\beta_{\bar{b}}$ have at least one coordinate that is either 0 or $L+1$).

Note how $[0, L+1]^h$ scales linearly with $L+1$ while $\pi_{\bar{b}}$ is a subspace and therefore scale-invariant. Hence, their intersection $\beta_{\bar{b}}(L)$ is an $h-1$-dimensional polytope that scales proportionally to the integer $L+1$, as required by Ehrhart's theorem [188]. The number $E_{\bar{b}}(L)$ of integer points in $\beta_{\bar{b}}(L)$ is then a polynomial in $L+1$ (and so L) of degree equal to the dimensionality of $\beta_{\bar{b}}(L)$, i.e., $h-1$. From Ehrhart-Macdonald's reciprocity theorem [189] we know that the number of integer points in the interior of $\beta_{\bar{b}}$ and thus in $\alpha_{\bar{b}}$ is $(-1)^{h-1} E_{\bar{b}}(-L)$, that is also a polynomial in L of degree $h-1$.

b) If two different assignments $\{\bar{b}'_l\}_{l=0}^{h-1}$ and $\{\bar{b}''_l\}_{l=0}^{h-1}$ are considered, then $\alpha_{\bar{b}'}(L) \cap \alpha_{\bar{b}''}(L) = [1, L]^h \cap \pi_{\bar{b}'} \cap \pi_{\bar{b}''}$. The same argument we used above tells us that the number of integer points in such an intersection is a polynomial in L of degree $h-2$ and, in general that the number of integer points in the intersection of any number of polytopes $\alpha_{\bar{b}}(L)$ is a polynomial of degree not larger than $h-1$.

c) The number of configurations of $\{u_l\}_{l=0}^{h-1}$ and $\{\bar{b}_l\}_{l=0}^{h-1}$ that satisfy (8.10) w.r.t. the above K is the number of integer points in the union of all possible polytopes $\alpha_{\bar{b}}$, i.e., $\bigcup_{\{\bar{b}_l\}_{l=0}^{h-1}} \alpha_{\bar{b}}(L)$. Such a number can be computed by the inclusion-exclusion principle that amounts to properly summing and subtracting the number of integer points in those polytopes and their various intersections. Since sum and subtraction of polynomials yield polynomials of non-increasing degree, we know that number is the evaluation of a polynomial $P_h(L)$ with degree not greater than $h-1$.

Let us finally write $P_h(L) = \sum_{j=0}^{h-1} p_j^h L^j$. In order to compute its coefficients p_j^h we may fix a binary configuration $\{\bar{b}_l\}_{l=0}^{h-1}$, count the points $\{u_l\}_{l=0}^{h-1} \in \mathbb{Z}_+^h$ for which (8.10) is verified by means of integer

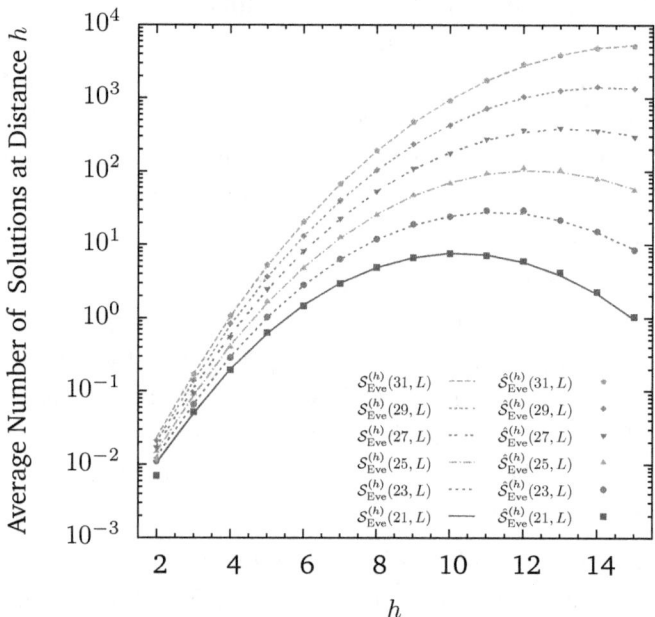

Figure 8.4: Empirical average number of solutions for Eve's KPA at Hamming distance h from the true one, compared to the theoretical approximation of (8.9) for $L = 10^4$ and $n = 21, 23, \ldots, 31$.

partition functions (that also have a polynomial expansion), and subtract the points in which $\{u_l\}_{l=0}^{h-1} \notin [1,L]^h$. By summation over all binary configurations, one can extract the coefficients associated with L^j for each h. Table 8.1 reports the result of this procedure as carried out by symbolic computation for $h \leq 15$. □

8.1.3 Class-Upgrade Known-Plaintext Attack

A KPA may also be attempted by Steve, a malicious second-class receiver aiming to improve its signal recovery performances with the intent of reaching the same quality of a first-class receiver. In this KPA, a partially correct encoding matrix $\mathbf{A}^{(0)}$ that differs from $\mathbf{A}^{(1)}$ by c entries is also known in addition to $\mathbf{x}$ and $\mathbf{y}$. With this prior, Steve may compute $\epsilon = \mathbf{y} - \mathbf{A}^{(0)}\mathbf{x} = \mathbf{\Delta A x}$ where $\mathbf{\Delta A} = \mathbf{A}^{(1)} - \mathbf{A}^{(0)}$ here is an unknown matrix with ternary-valued entries, i.e., $\mathbf{\Delta A} \in \{-2, 0, 2\}^{m \times n}$. Hence, Steve performs a KPA by searching for a set of ternary symbols $\{\Delta A_{j,l}\}_{l=0}^{n-1}$ such that each entry of ϵ,

$$\epsilon_j = \sum_{l=0}^{n-1} \Delta A_{j,l} x_l \qquad (8.11)$$

of which it is known *a priori* that $\Delta A_{j,l} \neq 0$ only in c cases. Moreover, to ease the solution of this problem and make it row-wise separable, we assume that Steve gains access to an even more accurate information, i.e., the exact number c_j of non-zero entries for each row $\mathbf{\Delta A}_j$ or equivalently the number of SSFs mapping $\mathbf{A}_j^{(0)}$ into the corresponding[4] $\mathbf{A}_j^{(1)}$. By assuming this, we may prove the equivalence between Steve's KPA to each row of $\mathbf{A}^{(1)}$ and a slightly adjusted SSP.

Problem 8.2 (γ-cardinality Subset-Sum Problem). Let $\{u_l\}_{l=0}^{n-1}, u_l \in \{1, \ldots, Q\}$, $\gamma \in \{1, \ldots, n\}$ and $v \in \mathbb{Z}_+$. We define γ-cardinality SSP (γ-SSP) the problem of assigning n binary variables $b_l \in \{0, 1\}$,

---

[4]Clearly, the total number of non-zero entries in $\mathbf{\Delta A}$ is $c = \sum_{j=0}^{m-1} c_j$.

$l = 0, \ldots, n-1$ so that

$$v = \sum_{l=0}^{n-1} b_l u_l \tag{8.12}$$

$$\gamma = \sum_{l=0}^{n-1} b_l \tag{8.13}$$

We define *solution* any $\{b_l\}_{l=0}^{n-1}$ verifying (8.12) and (8.13).

Again, a mapping of Steve's KPA to Problem 8.2 is easily obtained.

Proposition 8.2 *(Steve's Known-Plaintext Attack).* The KPA to $\mathbf{A}_j^{(1)}$ given $(\mathbf{x}, \mathbf{y}, \mathbf{A}^{(0)}, c_j)$, is equivalent to a γ-SSP where $\gamma = c_j$, $Q = 2L$, $u_l = -A_{j,l}^{(0)} x_l + L$, the variables

$$b_l = \frac{1}{2}\left(1 - \frac{\hat{A}_{j,l}^{(1)}}{A_{j,l}^{(0)}}\right)$$

and the sum

$$v = \frac{1}{2}\epsilon_j + L c_j$$

This SSP has a *true solution* $\{\bar{b}_l\}_{l=0}^{n-1}$ that is mapped to the row $\mathbf{A}_j^{(1)}$, and other *candidate solutions* that verify (8.12) and (8.13) but correspond to matrix rows $\hat{\mathbf{A}}_j^{(1)} \neq \mathbf{A}_j^{(1)}$.

We also define $(\mathbf{x}, \mathbf{y}, \mathbf{A}_j^{(0)}, \mathbf{A}_j^{(1)})$ a *problem instance*; Steve can therefore use the result of (8.12) to obtain the perturbation entries $\Delta A_{j,l} = -2A_{j,l}^{(0)} b_l$. The derivation of Proposition 8.2 is obtained as follows.

Proof of Proposition 2. In this case the attacker knows $(\mathbf{A}^{(0)}, \mathbf{x}, \mathbf{y})$, and is able to calculate $\epsilon = \mathbf{y} - \mathbf{A}^{(0)}\mathbf{x}$, i.e., $\epsilon_j = y_j - \sum_{l=0}^{n-1} A_{j,l}^{(0)} x_l = \sum_{l=0}^{n-1} \Delta A_{j,l} x_l$ where all the entries $\Delta A_{j,l}$ are unknown. For the j-th row, the attacker also knows there are c_j non-zero elements in $\Delta A_{j,l} = -2A_{j,l}^{(0)} b_l$ with $b_l \in \{0, 1\}$ binary variables that are 1 if the flipping occurred and 0 otherwise. Note that from the above information $c_j = \sum_{l=0}^{n-1} b_l$.

With this we define a set of even weights $D_l = -2A_{j,l}^{(0)}x_l \in \{-2L,\ldots,-2,0,2,\ldots,2L\}$ so the KPA is defined by satisfying the equalities

$$\epsilon_j = \sum_{l=0}^{n-1} D_l b_l \tag{8.14}$$

$$c_j = \sum_{l=0}^{n-1} b_l \tag{8.15}$$

To obtain a standard γ-SSP with positive weights and $\gamma = c_j$ we sum $2L$ to all D_l so (8.14) becomes $\epsilon_j + 2L\sum_{l=0}^{n-1} b_l = \sum_{l=0}^{n-1}(D_l + 2L)b_l$. Multiplying both sides by $1/2$ and using (8.15) yields $v = \frac{1}{2}\epsilon_j + Lc_j = \sum_{l=0}^{n-1} u_l b_l$ where $u_l = -A_{j,l}^{(0)}x_l + L \in \{0,\ldots,Q\}$. $Q = 2L$. Finally, we note the exclusion of $u_l = 0$ to facilitate the attack. □

In the following, we will denote with $r = c_j/n$ the row-density of perturbations. Since in [187] the γ-cardinality SSP case is obtained as an extension of the results on the unconstrained SSP, we obtain the following Theorem.

Theorem 8.3 (*Expected number of solutions for Steve's KPA*). *For large n, the expected number of candidate solutions of the KPA in Proposition 8.2, in which (i) all the coefficients $\{u_l\}_{l=0}^{n-1}$ are i.i.d. uniformly drawn from $\{1,\ldots,2L\}$, and (ii) the true solution $\{\bar{b}_l\}_{l=0}^{n-1}$ is drawn with equiprobable independent binary values, is*

$$S_{\text{Steve}}(n,L,r) \simeq \sqrt{\frac{3}{2}} \frac{r^{-1-nr}(1-r)^{-1-n(1-r)}}{2\pi nL} \tag{8.16}$$

The proof of Theorem 8.3 is reported in the next Section. The number of candidate solutions found by Steve's KPA is by many orders of magnitude smaller than Eve's KPA, the reason being that Steve requires much less information to achieve complete knowledge of the true encoding $\mathbf{A}^{(1)}$. In order to provide numerical evidence, we simulate Steve's KPA on a set of 50 randomly generated problem instances with row-density of perturbations $r = \{5/n, 10/n, 15/n\}$ for $n = \{20,\ldots,32\}$ and $L = 5 \cdot 10^3$; the problem is still formulated as binary programming

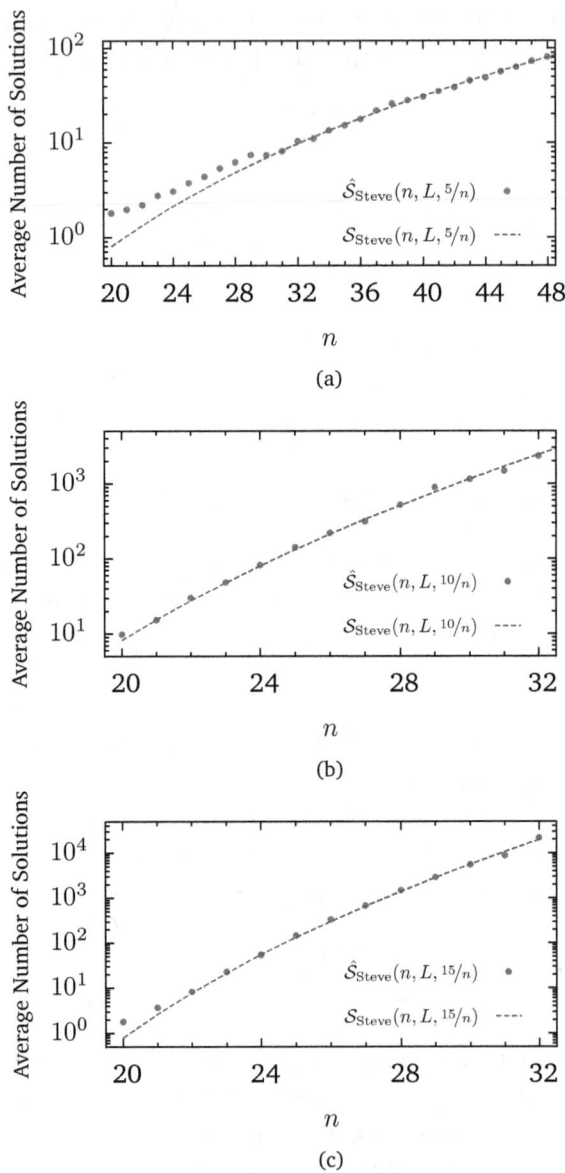

Figure 8.5: Empirical average number of solutions for Steve's KPA compared to the theoretical approximation of (8.16) for $L = 5 \cdot 10^3$ with row-density of perturbations $r = 5/n, 10/n, 15/n$.

in CPLEX, albeit with the additional equality constraint (8.15); the full solution pool can still be populated for the given dimensions[5].

The empirical average number of solutions $\hat{S}_{\text{Steve}}(n, L, r)$ reported in Fig. 8.5 is well predicted by the theoretical value in (8.16); note that this approximation is increasingly accurate for large n. Moreover, by resuming the previous example our $n = 64 \times 64$ pixel grayscale image quantised at $b_x = 8$ bit and encoded with two-class CS using $\Delta \mathbf{A}$ with $r = 0.03$ will have on-average $6.25 \cdot 10^{234}$ candidate solutions of indistinguishable quality.

The previous analysis hinges on a counting argument in a general setting, without any other prior assumption on the structure of $\mathbf{A}^{(1)}$ or $\Delta \mathbf{A}$. This class-upgrade KPA has been examined by assuming very accurate prior information on the number of perturbations per row, thus implying a best-case situation for the attacker. As we will show in the experiments of Section 8.3, these attacks yield no advantage in terms of recovery performances to unintended receivers.

A Proof of Theorem 8.3

The following proof draws again from the work of Sasamoto et al. [187] and is therefore similar in principle to that of Theorem 8.1, *i.e.*, it is merely an interface to existing results on the γ-SSP. It is worth noting that the proof draws on Definition 8.1.

Proof of Theorem 8.3. Assume $F_p(a, b)$ and $G_p(a, b)$ as in (8.5),(8.6). Define the normalised constraint $r = \frac{c_j}{n}$ and two quantities $a(\tau, r)$ and $b(\tau, r)$ that are the solutions of the following system of equalities

$$r = F_0(a, b)$$
$$\tau = F_1(a, b)$$

that are respectively equivalent to [187, (5.3-4)]. We also define

$$\mathcal{G}(\tau, r) = \begin{bmatrix} G_0(a\left(\tau, r\right), b\left(\tau, r\right)) & G_1(a\left(\tau, r\right), b\left(\tau, r\right)) \\ G_1(a\left(\tau, r\right), b\left(\tau, r\right)) & G_2(a\left(\tau, r\right), b\left(\tau, r\right)) \end{bmatrix}$$

---
[5] In the first case, full enumeration is still feasible in an acceptable computation time for $n = 48$.

With this, [187, (5.8-9)] prove that the number of solutions of a γ-SSP with integer coefficients $\{u_l\}_{l=0}^{n-1}$ uniformly distributed in $\{1,\ldots,Q\}, Q = 2L, \gamma = c_j$ is

$$\mathcal{S}_{\text{Steve}}(\tau, n, L, r) = \frac{e^{n(a(\tau,r)\tau - b(\tau,r))}}{4\pi nL\sqrt{\det(\mathcal{G}(\tau,r))}} e^{n\int_0^1 \log[1+e^{b(\tau,r)-a(\tau,r)\xi}]d\xi}$$

(8.17)

Using the same arguments as in the proof of Theorem 8.1, we average on τ and obtain an expression identical to (8.7) for the computation of $\mathbb{E}_\tau[\mathcal{S}_{\text{Steve}}(\tau, n, L, r)]$. Since $\mathcal{S}_{\text{Steve}}(\tau, n, L, r)$ has once again an approximately Gaussian profile in τ with a maximum in $\tau = \frac{r}{2}$ we approximate the expectation in τ,

$$\mathbb{E}_\tau[\mathcal{S}_{\text{Steve}}(\tau, n, L, r)] \simeq \mathcal{S}_{\text{Steve}}\left(\frac{r}{2}, n, L, r\right) \frac{1}{\sqrt{2}}$$

$$= \sqrt{\frac{3}{2}} \frac{r^{-1-n\rho}(1-r)^{-1-n(1-r)}}{2\pi n L} \quad (8.18)$$

by using the fact that $a\left(\frac{r}{2}, r\right) = 0$ and $b\left(\frac{r}{2}, r\right) = \log\left(\frac{r}{1-r}\right)$. □

8.2 Signal Recovery-Based Class-Upgrade Attacks

Class-upgrade attacks to two-class encryption by CS (Section 6.1.2) are closely related to a recovery problem that has attracted the attention of prior contributions, *i.e., sparse signal recovery under matrix uncertainty*, as was partly introduced in Chapter 5. In this case, we assume the perspective of Steve and let $\mathbf{A}^{(1)} = \mathbf{A}^{(0)} + \Delta\mathbf{A}$ be the encoding matrix, where $\mathbf{A}^{(0)}$ is known *a priori* and $\Delta\mathbf{A}$ is an unknown random SSF perturbation matrix. This qualifies as a class-upgrade known-ciphertext attack, as Steve is given $(\mathbf{y}, \mathbf{A}^{(0)})$ and no other information – if $\mathbf{x}$ was also provided, the best approach would still be the KPA in Proposition 8.2.

Steve's information could be paired with a sparsity prior on $\mathbf{x}$ to attempt the *joint recovery* of $\mathbf{x}$ and $\Delta\mathbf{A}$, eventually leading to a mere refinement of the estimate $\hat{\mathbf{x}}$ instead of an actual estimate of $\Delta\mathbf{A}$. Two main algorithms specifically address this problem setup for a generic $\mathbf{\Delta A}$, namely *Matrix Uncertainty-GAMP (MU-GAMP)* [160] and *Sparsity-cognisant Total Least-Squares (S-TLS)* [159].

Although appealing in principle, this joint recovery approach can be anticipated to fail for multiple reasons. First, this attack is intrinsically harder than Steve's KPA in that the true plaintext x here is unknown. Whatever $\mathbf{\Delta A}$ is a candidate solution to Steve's KPA given x, it will also be possible solution of joint recovery with the same x. Since we know from Section 8.1.3 that Steve's KPA typically has a huge number of indistinguishable and equally-sparse candidate solutions, at least as many will verify the joint recovery problem when the plaintext is unknown. Hence, this approach has negligible odds of yielding more information on $\mathbf{\Delta A}$ than Steve's KPA.

Furthermore, note that joint recovery amounts to solving $\mathbf{y} = \mathbf{A}^{(0)}\mathbf{x} + \mathbf{\Delta A}\mathbf{x}$ with $\mathbf{\Delta A}$ and x unknown, that is clearly a non-linear equality involving non-convex/non-concave operators; in general, this is a hard problem that can only be solved in a relaxed form (as, in fact, does S-TLS).

The aforementioned algorithms are indeed able to compensate matrix uncertainties when $\mathbf{\Delta A}$ depends on a low-dimensional, deterministic set of parameters. However such a model does not apply to two-class encryption by CS: even if $\mathbf{\Delta A}$ is c-sparse, it has no deterministic structure to leverage in the attack – to make it so, one would need to know the exact set $C^{(0)}$ of c index pairs at which the sign-flipping randomly occurred, which by itself entails a combinatorial search.

In fact, $\mathbf{\Delta A}$ is *uniform* in the sense of [160] since it is a realisation of a RME with i.i.d. entries having zero-mean and bounded variance. Hence, we expect the accuracy of the estimate $\hat{\mathbf{x}}$ with joint recovery (both using S-TLS and MU-GAMP) to agree with the uniform matrix uncertainty case of [160], where negligible improvement is shown w.r.t.. the (standard, non-joint) recovery algorithm GAMP [105]. The advocated reason is that the perturbation noise $\epsilon = \mathbf{\Delta A}\mathbf{x}$ is asymptotically Gaussian for a given x [160, Proposition 2.1]; thus, it is reasonable that a suitably-tuned application of GAMP attains near-optimal performances.

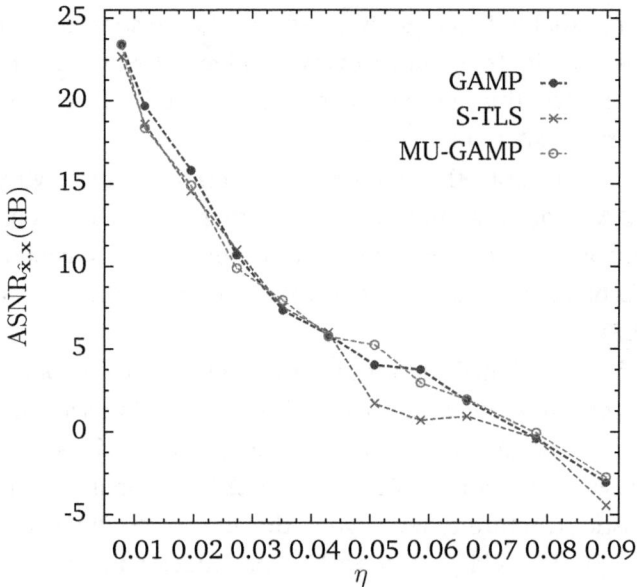

Figure 8.6: $\mathrm{ASNR}_{\hat{x},x}$ performances of a class-upgrade known-ciphertext attack using signal recovery under matrix uncertainty algorithms.

We now provide empirical evidence on the ineffectiveness of joint recovery as a class-upgrade attack for finite n, m and sparsity k. As an example, we let $n = 256$, $m = 128$, $k = 20$ and $\eta = \frac{c}{mn} \in [0.005, 0.1]$ and generate 100 random instances of $\mathbf{x} = \mathbf{D}\mathbf{s}$ with $\mathbf{x}$ being k-sparse w.r.t. a randomly chosen orthonormal basis $\mathbf{D}$. For each η, we also generate 100 pairs of matrices $(\mathbf{A}^{(0)}, \mathbf{A}^{(1)})$ related as (6.2) and encode $\mathbf{x}$ by $\mathbf{y} = \mathbf{A}^{(1)}\mathbf{x}$.

Signal recovery is performed by MU-GAMP, S-TLS and GAMP. To maximise their performances, each of the algorithms is "genie"-tuned to reveal the exact value of the required features of $\mathbf{x}$. In particular, MU-GAMP and GAMP are provided with an i.i.d. Bernoulli-Gaussian sparsity-enforcing signal model [105, 170] having the exact mean, variance and sparsity level of the instances s. As far as the perturbation $\mathbf{\Delta A}$ is concerned, MU-GAMP is given the PMF of its i.i.d. entries. On the other hand, GAMP is initialised with the noise variance of $\epsilon = \mathbf{\Delta A}\mathbf{x}$, that is assumed as AWGN. S-TLS is run in its locally-optimal, polynomial-time version [159, Section IV-B] and fine-tuned w.r.t. its regularisation parameter as η varies.

Since the typically very low accuracy of the recovered $\mathbf{\Delta A}$ is not as relevant to a class-upgrade attack as improving the estimate of $\hat{\mathbf{x}}$, we here focus on measuring the usual $\text{ASNR}_{\hat{\mathbf{x}}, \mathbf{x}}$, as reported in Fig. 8.6. The standard deviation from the average is less than $1.71\,\text{dB}$ in all the reported curves. The maximum $\text{ASNR}_{\hat{\mathbf{x}}, \mathbf{x}}$ performance gap between GAMP and MU-GAMP is $1.22\,\text{dB}$ while S-TLS attains generally lower performances for high values of η. These observed performances confirm what is also found in [160], *i.e.*, that GAMP, MU-GAMP and S-TLS substantially attain the same performances under uniform matrix uncertainty. As expected, class-upgrade attacks based on joint recovery are ineffective even for finite n and m, since GAMP under the same conditions is the reference case adopted in Section 6.3 for the design of two-class encryption by CS.

8.3 Performance Evaluation of Practical Known-Plaintext Attacks

In this Section we exemplify KPAs in a common framework which entails the following procedure. When Eve is performing a KPA as in Section 8.1, it knows a single plaintext-ciphertext pair $(\mathbf{x}', \mathbf{y}')$ and attacks a matrix $\mathbf{A}^{(1)}$ row-by-row; we here infer each row $\mathbf{A}_j^{(1)}$ by generating random instances of an i.i.d. RBE matrix until a chosen number of candidate rows $\hat{\mathbf{A}}_j^{(1)}$ that verify $y'_j = \hat{\mathbf{A}}_j^{(1)} \mathbf{x}'$ has been found. Thus, the inferred $\hat{\mathbf{A}}^{(1)}$ is actually composed by collecting the outputs of m random searches. This approach is preferable to solving Eve's KPA by means of linear programming as in Section 8.1 for two reasons.

Firstly, it is known from Theorem 8.1 that the expected number of solutions is very large and thus the probability of success of a random search is far from being negligible, while its computational cost is relatively low.

Secondly, the theoretical conditions of Section 1.3.2 that guarantee when $\mathbf{x}'$ can be retrieved from $\mathbf{y}'$ despite the dimensionality reduction are applicable when $\mathbf{A}^{(1)}$ is drawn from the RBE. On the contrary, the CPLEX integer programming solver explores solutions in a systematic way, and while crucial in the enumeration of *all* candidate solutions as in Section 8.1 (with computational cost growing exponentially in n) it tends to generate them in an ordered fashion. When only some of these solutions are considered (as obliged when n is large) this results in sets of $\hat{\mathbf{A}}_j^{(1)}$ that could be very distant from $\mathbf{A}_j^{(1)}$.

To test the obtained guess $\hat{\mathbf{A}}^{(1)}$, Eve may then *pretend* to ignore $\mathbf{x}'$ and recover its approximation $\hat{\mathbf{x}}'$ from $(\mathbf{y}', \hat{\mathbf{A}}^{(1)})$ by using a high-performance signal recovery algorithm such as GAMP [105] optimally tuned as in Section 8.2. This sets the $\text{SNR}_{\hat{\mathbf{x}}', \mathbf{x}}$ level which is adopted as a quality indicator for $\hat{\mathbf{A}}^{(1)}$.

Then, Eve attempts signal recovery from a second ciphertext $\mathbf{y}'' = \mathbf{A}^{(1)} \mathbf{x}''$ where the plaintext $\mathbf{x}''$ is unknown, *i.e.*, as if somehow $\mathbf{A}^{(1)}$ was reused twice. In this case, and if Eve's KPA was successful in retrieving $\hat{\mathbf{A}}^{(1)}$, the recovery $\hat{\mathbf{x}}''$ obtained by means of GAMP would yield a new $\text{SNR}_{\hat{\mathbf{x}}'', \mathbf{x}''} \approx \text{SNR}_{\hat{\mathbf{x}}', \mathbf{x}'}$. To remark what is shown below, we evaluate how the $(\mathrm{SNR}_{\hat{\mathbf{x}}',\mathbf{x}'}, \mathrm{SNR}_{\hat{\mathbf{x}}'',\mathbf{x}''})$ pairs are distributed w.r.t. fixed plaintexts $\mathbf{x}', \mathbf{x}''$ encoded with the same $\mathbf{A}^{(1)}$ and candidate solutions $\hat{\mathbf{A}}^{(1)}$ are considered in the decoding; if Eve is successful, an $\mathrm{SNR}_{\hat{\mathbf{x}}'',\mathbf{x}''}$ compatible with $\mathrm{SNR}_{\hat{\mathbf{x}}',\mathbf{x}'}$ must be observed.

The examples of class-upgrade KPAs follow the same procedure as those performed by Eve, with the exception that Steve generates the rows of $\hat{\mathbf{A}}^{(1)}$ by random search of the index set $C_j^{(0)}$ that maps the known $\mathbf{A}_j^{(1)}$ to $\hat{\mathbf{A}}_j^{(1)}$ that verifies $y'_j = \hat{\mathbf{A}}_j^{(1)} \mathbf{x}'$. Coherently with the theoretical setting of Section 8.1.3, we also assume that Steve knows exactly c_j entries are flipped in each row. Repeating this search for m rows provides the candidate solutions $\hat{\mathbf{A}}^{(1)}$, of which we will study how the corresponding $(\mathrm{SNR}_{\hat{\mathbf{x}}',\mathbf{x}'}, \mathrm{SNR}_{\hat{\mathbf{x}}'',\mathbf{x}''})$ pairs are distributed as mentioned above.

8.3.1 Electrocardiographic Signals

We now consider ECG signals in the same conditions of Section 6.3, focusing on two windows $\mathbf{x}', \mathbf{x}''$ of $n = 256$ samples quantised with $b_\mathbf{x} = 12$ bit; these correspond to the measurement vectors $\mathbf{y}', \mathbf{y}''$ of dimensionality $m = 90$. Signal recovery is allowed by the sparsity level of the windowed signal when decomposed with $\mathbf{D}$ chosen as a Symmlet-6 orthonormal DWT [52].

We generate 2000 candidate solutions for both Eve and Steve's KPA that correspond to the recovery performances reported in Fig. 8.7. While both malicious users are able to reconstruct the known plaintext $\mathbf{x}'$ with a relatively high[6] average $\mathrm{SNR}_{\hat{\mathbf{x}}',\mathbf{x}'} \approx 25\,\mathrm{dB}$, on the second window of samples $\mathbf{x}''$ the eavesdropper achieves an average $\mathrm{SNR}_{\hat{\mathbf{x}}'',\mathbf{x}''} \approx -0.20\,\mathrm{dB}$ (Fig. 8.7a), whereas the second-class decoder achieves an average $\mathrm{SNR}_{\hat{\mathbf{x}}'',\mathbf{x}''} \approx 12.15\,\mathrm{dB}$ (Fig. 8.7b) when the two-class encryption scheme is set to a sign flipping density $\eta = c/mn = 0.03$ between $\mathbf{A}^{(1)}$ and $\mathbf{A}^{(1)}$. In this case, the nominal second-class $\mathrm{RSNR} = 11.08\,\mathrm{dB}$ when reconstructing $\mathbf{x}''$ from $\mathbf{y}''$ with $\mathbf{A}^{(1)}$, while the correlation coefficient between $\mathrm{SNR}_{\hat{\mathbf{x}}',\mathbf{x}'}$ and $\mathrm{SNR}_{\hat{\mathbf{x}}'',\mathbf{x}''}$ is 0.0140; these figures clearly highlight the ineffectiveness of KPAs at inferring

---

[6] Their KPAs indeed yield solutions to $\mathbf{y}' = \hat{\mathbf{A}}^{(1)} \mathbf{x}'$.

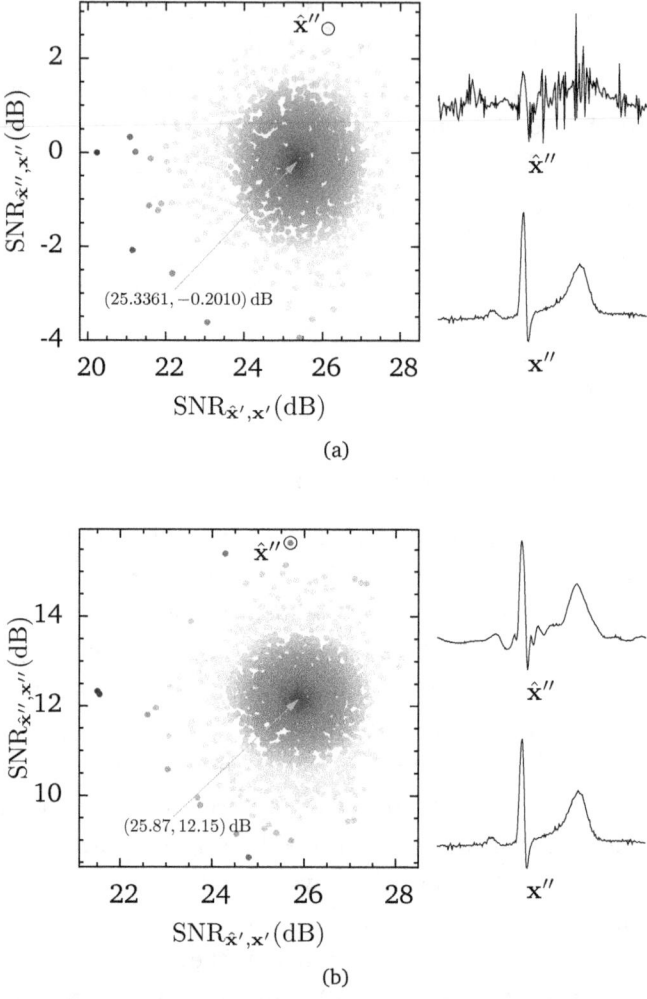

Figure 8.7: Effectiveness of (a) Eve and (b) Steve's KPA in recovering a hidden ECG. Each point is a guess of the encoding matrix $\mathbf{A}^{(1)}$ whose quality is assessed by decoding the ciphertext y' corresponding to the known plaintext x' ($\text{SNR}_{\hat{\mathbf{x}}',\mathbf{x}'}$) and by decoding a new ciphertext y'' ($\text{SNR}_{\hat{\mathbf{x}}'',\mathbf{x}''}$). The Euclidean distance from the average ($\text{SNR}_{\hat{\mathbf{x}}',\mathbf{x}'}, \text{SNR}_{\hat{\mathbf{x}}'',\mathbf{x}''}$) is highlighted by colour gradient.

$\mathbf{A}^{(1)}$ in this case. This is also confirmed by the perceptual quality of $\hat{\mathbf{x}}''$ corresponding to the maximum $\mathrm{SNR}_{\hat{\mathbf{x}}'',\mathbf{x}''}$ highlighted in Fig. 8.7.

8.3.2 Sensitive Text in Images

In this example we consider the same test images used in Section 6.3, *i.e.*, 640×512 pixel grayscale images of people holding a printed identification text concealed by means of two-class encryption. To reduce the computational burden of KPAs we assume a block size of 64×64 pixel, $b_\mathbf{x} = 8$ bit per pixel, and encode the resulting $n = 4096$ pixels into $m = 2048$ measurements. Signal recovery is performed by assuming the blocks have a sparse representation on a 2D Daubechies-4 orthonormal DWT [52]. Two-class encryption is applied on the blocks containing printed text: we choose two adjacent blocks $\mathbf{x}', \mathbf{x}''$ containing some letters and encoded with the same $\mathbf{A}^{(1)}$; in this case, the second-class decoder nominally achieves RSNR $= 12.57$ dB without attempting class-upgrade due to the flipping of $c = 251658$ entries (corresponding to a perturbation density $\eta = 0.03$) in the encoding matrix.

In order to test Eve and Steve's KPA we randomly generate 2000 solutions for the j-th row of the encoding given $\mathbf{x}', \mathbf{y}'$: it is worth noting that while in the previous case the signal dimensionality is sufficiently small to produce a solution set in less than two minutes, in this case generating 2000 different solutions for a single row may take up to several hours for particularly hard instances. By using these candidate solutions to find $\hat{\mathbf{x}}', \hat{\mathbf{x}}''$ we obtain the results of Fig. 8.8: while both attackers attain an average $\mathrm{SNR}_{\hat{\mathbf{x}}',\mathbf{x}'} \approx 33$ dB on $\mathbf{x}'$, Eve is only capable of reconstructing $\mathbf{x}''$ with an average $\mathrm{SNR}_{\hat{\mathbf{x}}'',\mathbf{x}''} \approx 0.14$ dB where Steve reaches an average $\mathrm{SNR}_{\hat{\mathbf{x}}'',\mathbf{x}''} \approx 12.80$ dB with $\eta = 0.03$.

Note also that, although lucky guesses exist with $\mathrm{SNR}_{\hat{\mathbf{x}}'',\mathbf{x}''} > 12.57$ dB, it is impossible to identify them by looking at $\mathrm{SNR}_{\hat{\mathbf{x}}',\mathbf{x}'}$ since the correlation coefficient between $\mathrm{SNR}_{\hat{\mathbf{x}}',\mathbf{x}'}$ and $\mathrm{SNR}_{\hat{\mathbf{x}}'',\mathbf{x}''}$ is -0.0041. Thus, Steve cannot rely on observing the $\mathrm{SNR}_{\hat{\mathbf{x}}',\mathbf{x}'}$ to choose the best performing solution $\hat{\mathbf{A}}^{(1)}$, so both Eve and Steve's KPAs are inconclusive. As a further perceptual evidence of this, the best

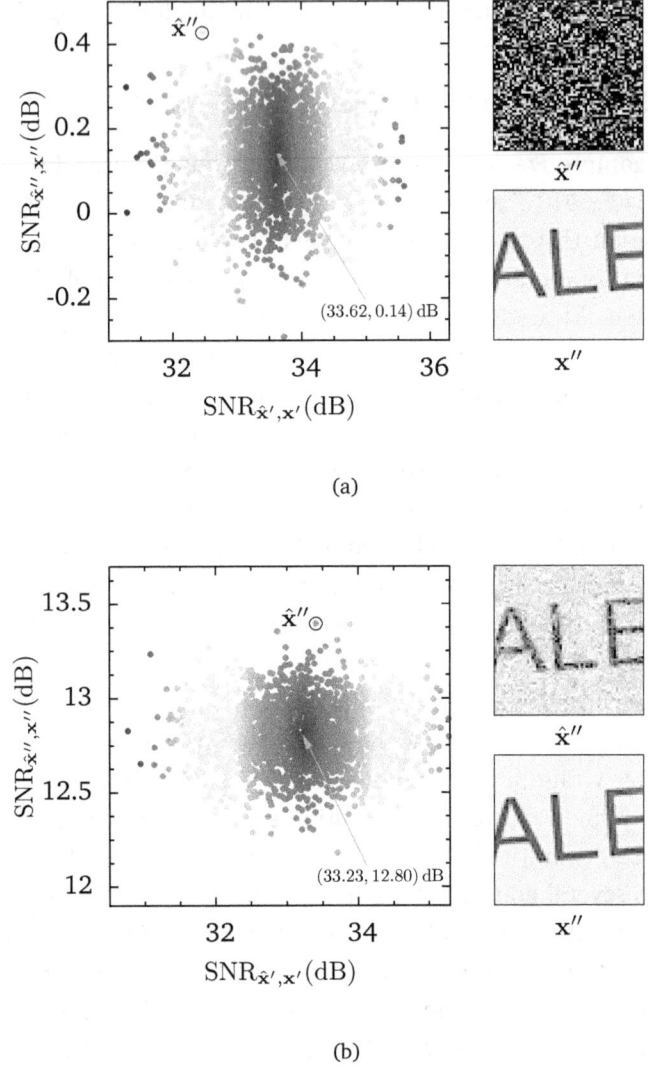

Figure 8.8: Effectiveness of (a) Eve and (b) Steve's KPA in recovering hidden image blocks. Each point is a guess of the encoding matrix $\mathbf{A}^{(1)}$ whose quality is assessed by decoding the ciphertext y' corresponding to the known plaintext x' ($\text{SNR}_{\hat{\mathbf{x}}',\mathbf{x}'}$) and by decoding a new ciphertext y'' ($\text{SNR}_{\hat{\mathbf{x}}'',\mathbf{x}''}$). The Euclidean distance from the average ($\text{SNR}_{\hat{\mathbf{x}}',\mathbf{x}'}, \text{SNR}_{\hat{\mathbf{x}}'',\mathbf{x}''}$) is highlighted by colour gradient.

recoveries according to the $\mathrm{SNR}_{\hat{x}'',x''}$ are reported in Fig. 8.8.

Summary

- We have analysed KPAs as they may be carried out on standard CS schemes with RBE encoding matrices as well as on the particular multiclass protocol developed in Chapter 6. In particular, the analysis was carried out from the two perspectives of an eavesdropper and a second-class user trying to guess the true encoding matrix.

- A theoretical approach to KPAs shows how they can be mapped into two versions of the SSP, with the aim of counting the candidate encoding matrices that match a given plaintext-ciphertext pair.

- In the eavesdropper case we have found that for each row the expected number grows as $\mathcal{O}(2^n\, n^{-\frac{1}{2}})$; thus, finding the true solution among such huge sets is infeasible.

  A further study of the candidate solutions' Hamming distance from the true one showed that, as the dimensionality n increases, the expected number of solutions close to the true one is only a small fraction of the solution set.

- As for the second-class user we have shown that depending on the available information on the true encoding matrix, the expected number of solutions is significantly smaller, yet sufficiently high for large n to reassure that a second-class user will not be able to perform class-upgrade.

- Signal recovery-based class-upgrade attacks were attempted by using recovery algorithms that account for matrix perturbations; these were here shown to yield almost identical performances to those of a standard decoding algorithm because of the random nature of the SSF perturbation matrix.

- Finally, we showed practical KPAs on real-world signals such as ECG traces and images by running a random search for a solution set corresponding to realistic plaintext-ciphertext pairs, and afterwards tested whether any of the returned candidate solutions could lead to finding the true encoding matrix by testing them on a successive ciphertext encoded with the same matrix.

In all the observed cases, we have found that the decoding performances match the $\text{ASNR}_{\hat{\mathbf{x}},\mathbf{x}}$ level prescribed by the multi-class encryption protocol, *i.e.*, both malicious users are unable to successfully decode other plaintexts with significant and stable quality improvements w.r.t. their available prior information.

III

A Multispectral Compressive Imager by Random Convolution

A MULTISPECTRAL COMPRESSIVE IMAGER BY RANDOM CONVOLUTION

lettrine[nindent=0.25em]Computer vision applications entail the acquisition of high-resolution scenes comprised of objects, shapes and landscapes which are identified by simple morphological features defining their information content. These scenes are very often accurately represented by means of the techniques described in Section 1.1.1; in fact, lossy digital image compression schemes [131] explicitly leverage the principle of compressibility (as in Definition 1.4), enabling efficient and compact storage of compressed images with negligible perceptual quality losses w.r.t. uncompressed, raw frames acquired from imaging sensors. Regardless of the redundancy indicated by the existence of a low-dimensional model, state-of-the-art imaging sensors rely on increasingly large pixel counts to acquire raw frames.

This quest for high-resolution sensors becomes even more relevant when MS imaging is considered. This imaging technique is commonly implemented by spatially or spectrally multiplexing the acquisition of parts of a MS data volume on a standard imaging sensor. To do so, both spatial [63, 190] and spectral [191] multiplexing architectures exist[1], yet they require accurate dispersive optics (such as prisms, gratings or tunable optical filters) to separate the spectral components of a scene. In addition, precise mechanical scanning elements are often needed to

---

[1]Reviews on imaging spectrometry may be found in [192, 193].

complete the acquisition, consequently causing motion artefacts in the presence of rapid temporal variations in the scene.

Recent developments [194, 195] have shown that MS image acquisition can be performed in a more agile fashion by using *Fabry-Pérot (FP)* spectral band-pass filters [196] deposited directly above a standard CMOS imaging sensor (*i.e.*, as in Fig. 9.1), paired with a lens array that replicates the MS cube on the FP-filtered imaging sensor. The introduction of this technology allows the removal of dispersive optics and scanning elements from the optical system, thus increasing its compactness and reducing the acquisition time required to sample a single cube to that of acquiring a single frame (*i.e.*, a *snapshot*). The price of this major simplification is a limit in the spatial resolution of the data volume, since the total pixel count of the sensor is partitioned by the number of wavelengths being acquired at each snapshot. Thus, snapshot MS imagers are generally more limited in resolution than their scanning counterparts, and would therefore benefit the most from the use of sensing techniques that increase the spatial-domain data volume resolution.

In this Chapter we document some advancements and findings regarding the development, implementation and experimental testing of a MS imager based on CS and the above FP-filtered imaging sensors. Our imager is based on the principle of *random convolution*, which provides a convenient sensing operator for CS [30]. This principle was recently applied by Björklund and Magli [32] to devise a panchromatic single-snapshot imaging scheme based on CS. Starting from the latter, we here extend their design to the task of MS imaging with FP-filtered sensors and show by simulation how single- and multi-snapshot acquisition can be flexibly used to sample MS cubes with accuracy depending on their complexity. Moreover, we here bring this optical design into a prototype that uses some standard optical components, a programmable light modulator and a FP-filtered sensor array.

We anticipate that our scheme applies random convolution independently on each slice of the data volume, and therefore introduces no CS in the wavelength domain; this choice favours lower system-level

Figure 9.1: A Fabry-Pérot-filtered sensor array; image courtesy of IMEC, Belgium.

complexity and the removal of complex dispersive optical elements (contrarily to the MS imaging schemes in [197, 198]).

A fundamental role will also be played by the discrepancy between the ideal sensing operator model and its optical implementation, that will affect the sensing scheme and require at least a *Point Spread Function (PSF)* estimation step to refine the ideal model. At the present state, this discrepancy limits the quality of the recovery results obtained from both a panchromatic compressive imager based on random convolution and its MS companion; yet, it is on this estimation and calibration step that future improvements will focus.

9.1 Compressive Imaging by Random Convolution

Compressed Sensing as illustrated in Chapter 1 can be implemented in the optical domain by a variety of schemes devised in recent contributions [20, 197, 199–201] (see also the tutorial in [151], and references therein). Some of these specifically address the choice of a *random convolution* sensing operator [30, 199, 202] which is the core of the MS imager we here develop. The appeal of such a scheme is in the parallelism provided by the convolution operation in the optical domain, as opposed to the operation mode of the *single-pixel camera* [20] where the acquisition of m measurements is completely serialised in time. In particular, while optical convolution schemes may be devised by using the Fourier-transforming property of lenses [203, Section 5.2] the main novelty of the scheme explored by Björklund and Magli [32], on which our MS imager is based, is that it uses an *out-of-focus coded aperture array*. This and other concepts will be cleared out in this Section as a background to the simulation and implementation of our MS imager.

9.1.1 Coded Aperture Imaging

The convolution operator $*$ is a fundamental tool in optical and digital image processing, as it implements elementary linear filtering that allows the extraction of relevant spatial information from the scene being imaged. In the continuous, two-dimensional spatial domain it is defined as

$$y(u,v) = [x * h](u,v) = \int_{\mathbb{R}^2} x(u-\xi, v-\zeta) h(\xi, \zeta) \mathrm{d}\xi \mathrm{d}\zeta \qquad (9.1)$$

where $h(u,v)$ is the *convolution kernel* or PSF and $x(u,v)$ is the light intensity of the scene being imaged. This operator allows the description of linear shift-invariant filters in the optical domain[2].

In particular, (9.1) models image formation schemes that use *spatial filters* as optical processing elements, *e.g.*, apertures or aperture arrays. The simplest example of such a scheme is the *pinhole camera*, in which images of a scene are produced on an observation plane by blocking the incident light rays with an opaque screen, with the exception of a very small aperture (*i.e.*, a *pinhole*). The passage of light rays propagating from the scene through this aperture can be modelled as a convolution of the scene with a kernel $h(u,v)$, which is commonly chosen as a 2D Gaussian profile of unit amplitude and width at half-maximum depending on the radius of the aperture. Clearly, the smaller the radius, the lower the light intensity transmitted from the source to the observation plane; however, the larger the radius, the blurrier

---

[2] More generally, the theory of Fourier optics describes wave propagation by means of Fourier transforms and convolution operators; see [203] for an in-depth introduction to the topic.

the image obtained on the observation plane, indicating a trade-off between light transmittance[3] and resolution.

Coded aperture imaging was devised to overcome this transmittance decrement, and is a valuable technique in cases where image formation by pinholes is more feasible than with standard lenses. Coded apertures are opaque screens with arrays of pinholes; these process incident light rays by forming shifted-and-superimposed copies of a scene on an observation plane (for a review of the topic, see [204]).

Clearly, this multiplicity augments light transmittance proportionally to the number of pinholes, while making the measurements *indirect*, *i.e.*, the image of a scene is not *formed* as in standard imaging systems, but rather *recovered* from the convolved measurements by means of a suitable algorithm which essentially inverts the effect of a sensing operator $\mathcal{A}$ (similarly to the principle of CS in Fig. 1.2).

The optical operation performed by a pinhole array can be modelled by a convolution kernel such as the ideal[4] Dirac pulse train $h(u,v) = \sum_{j=0}^{n_p-1} \delta(u-u_j, v-v_j)$ where $\delta(u,v)$ is the 2D Dirac delta and (u_j, v_j) denote the centre coordinates of each of n_p pinholes. In this way, the convolved image at the observation plane is a superposition of shifted and spatially-filtered versions of the original object.

A *deconvolution* algorithm will then operate on a discrete model of the optical processing chain, *i.e.*, $\mathbf{y} = \mathbf{M} * \mathbf{x}$. Here we let $\mathbf{M} \in \{0,1\}^{s_x \times s_y}$ represent a uniform grid of square apertures, that is a discretisation of the convolution kernel of the aperture array; $\mathbf{x} \in \mathbb{R}^{n_x \times n_y}$ is the Nyquist-rate representation of a scene and $*$ denotes a discrete 2D convolution. Depending on the boundary conditions on $\mathbf{x}$, the linear convolution $\mathbf{y} \in \mathbb{R}^{n_x+s_x-1 \times n_y+s_y-1}$ is obtained under zero boundary values, whereas circular convolution is obtained when cyclic boundary values are assumed (*i.e.*, as by the 2D DFT) yielding $\mathbf{y} \in \mathbb{R}^{s_x \times s_y}$ where $s_x \geq n_x, s_y \geq n_y$.

In addition, special kernels exist for which the deconvolution algorithm is replaced by a circular convolution: *Uniformly Redundant*

---

[3]The ratio of transmitted light intensity from the source to the observation plane.
[4]A more realistic model can be derived by using a similar train of Gaussian-shaped apertures.

Arrays (URAs) [205] are designed on the principle that a complementary kernel $\mathbf{G} \in \{-1,1\}^{s_x \times s_y}$ exists, for which the convolution $\forall (j,l) \in \{0,\ldots,s_x-1\} \times \{0,\ldots,s_y-1\}$, $(\mathbf{G} * \mathbf{M})_{j,l} = \delta(j,l)$. Simply put, the cross-correlation of $\mathbf{G}$ and $\mathbf{M}$ is a 2D Kronecker delta; such apertures can be constructed for a prime number $s_x = s_y$. Using this scheme, deconvolving $\mathbf{x}$ is as simple as applying another circular convolution: when the measurements $\mathbf{y} = \mathbf{M} * \mathbf{x} + \boldsymbol{\nu}$, with $\boldsymbol{\nu} \in \mathbb{R}^{s_x \times s_y}$ denoting, *e.g.*, AWGN noise, one may deconvolve $\hat{\mathbf{x}} = \mathbf{G} * \mathbf{y} = \mathbf{x} + \mathbf{G} * \boldsymbol{\nu}$. A residual noise term is therefore present in (and may eventually be accentuated by) this deconvolution process.

More complex methods for the solution of a linear inverse problem such as those outlined in Section 1.3 may also be applied to recover $\mathbf{x}$ given $(\mathbf{y}, \mathbf{M})$, although the measurements $\mathbf{y}$ here are not undersampled w.r.t. the dimensionality of $\mathbf{x}$. URA coded apertures will find application as a calibration tool for the proposed imager.

9.1.2 Compressed Sensing by Random Convolution

The connection between coded aperture imaging and CS is quite immediate by considering as a sensing operator the matrix $\mathbf{A} \in \mathbb{R}^{m \times n}$ that corresponds to a 2D cyclic or linear convolution by some $\mathbf{M} \in \mathbb{R}^{s_x \times s_y}$ followed by a further operation that reduces the convolution elements' dimensionality. Starting from a full-size linear convolution $\mathbf{M} * \mathbf{x}$ of dimensions $n_x + s_x - 1 \times n_y + s_y - 1$ in a 2D domain, by a suitable selection of its elements[5] we obtain the sensing operator

$$\mathbf{y} = \mathcal{A}(\mathbf{x}) = \mathbf{P}^{\Omega} \left[\text{vec}\,(\mathbf{M} * \mathbf{x}) \right] \in \mathbb{R}^m \qquad (9.2)$$

where

$$\mathbf{P}^{\Omega} \in \{0,1\}^{m \times q}, \; q = (n_x + s_x - 1)(n_y + s_y - 1)$$

is a randomly chosen selection matrix (as used in defining the PFE in Section 1.2.3), *i.e.*, $|\Omega| = m$; note that the measurements are denoted in vector form, but may be equivalently mapped to $\mathbf{y} \in \mathbb{R}^{m_x \times m_y}$, $m =$

---

[5] While the full-size linear convolution has dimensions $n_x + s_x - 1 \times n_y + s_y - 1$, assuming $s_x > n_x, s_y > n_y$ only $s_x - n_x + 1 \times s_y - n_y + 1$ of these elements are obtained with $\mathbf{M}$ and $\mathbf{x}$ completely overlapping.

$m_x m_y$. Since a full-size linear convolution has a matrix representation $\overline{\mathbf{M}}$ comprised of Toeplitz blocks [206, Section 2.8], verifying that $\mathcal{A}$ is a valid sensing operator amounts to evaluating whether $\mathbf{A} = \mathbf{P}^\Omega \overline{\mathbf{M}}$ has the RIP. Remarkably, the answer is positive: Rauhut [81, Theorem 1.1] showed that a partial circulant Toeplitz matrix $\mathbf{A}$, as originated by (9.2) with $\mathbf{M}$ drawn from the RBE and $\mathbf{P}^\Omega$ an m-dimensional selection matrix, is still endowed with the RIP; its k-RIC is so that $\mathbb{P}[\delta_k \leq \delta] \simeq 1$ provided that $m \geq \overline{m}$, with

$$\overline{m} = \mathcal{O}\left(\max\left\{\delta^{-1} k^{\frac{3}{2}} (\log n)^{\frac{3}{2}}, \delta^{-2} k (\log k)^2 (\log n)^2\right\}\right) \quad (9.3)$$

This same observation holds equivalently for Toeplitz-block sensing matrices (*i.e.*, 2D convolutions). Regrettably, a direct comparison of (9.3) with the behaviour of RsGE sensing matrices in (1.14) shows how (9.2) is strongly sub-optimal w.r.t. them; yet, aside from the undersampling implied by $\mathbf{P}^\Omega$, $\mathcal{A}$ is a *simple* convolution by an RBE matrix $\mathbf{M}$.

In more detail, since $\mathbf{M}$ will be implemented by a coded aperture in the optical domain, its negative values cannot be physically represented. Thus, the *effective* coded aperture pattern will be $\mathbf{M}^+ \in \{0,1\}^{s_x \times s_y}$, $\mathbf{M}^+ = \frac{1}{2}(\mathbf{M} + \mathbf{1}_{s_x \times s_y})$. Hence, by the linearity of (9.2), we will map the *positive measurements* $\mathbf{y}^+ = \mathbf{P}^\Omega \text{vec}\,(\mathbf{M}^+ * \mathbf{x})$ to $\mathbf{y}$ by considering[6]:

1. the difference between positive and negative measurements, *i.e.*,

$$\mathbf{y} = \mathbf{y}^+ - \mathbf{y}^- = \mathbf{P}^\Omega \text{vec}\,((\mathbf{M}^+ - \mathbf{M}^-) * \mathbf{x}), \, \mathbf{M}^- = \mathbf{1}_{s_x \times s_y} - \mathbf{M}^+ \quad (9.4)$$

   if it is conveniently implemented in the analog domain;

2. the difference between positive measurements and a reference, *i.e.*, $\mathbf{y} = 2\mathbf{y}^+ - \mathbf{y}^w$ with $\mathbf{y}^w = \mathbf{P}^\Omega \text{vec}\,(\mathbf{1}_{s_x \times s_y} * \mathbf{x})$, where the additional reference measurement is obtained with a fully open coded aperture $\mathbf{M} = \mathbf{1}_{s_x \times s_y}$.

---

[6]It is also worth noting that both the proposed strategies neglect the impact of noise, that will differ in the acquisition of each instance of $\mathbf{y}^+, \mathbf{y}^-$ and $\mathbf{y}^w$.

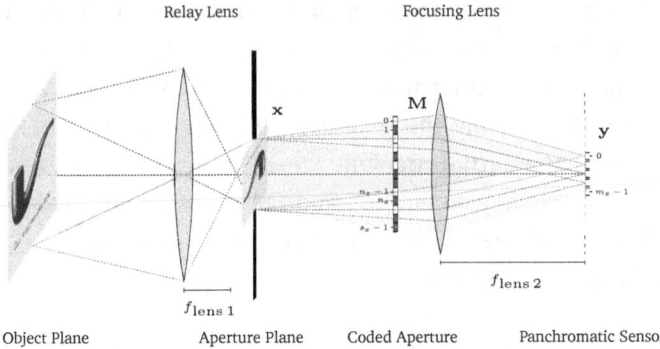

Figure 9.2: Optical scheme (sagittal section) of a panchromatic compressive imager by out-of-focus random convolution. The rays traced before the coded aperture are coloured depending on their angle w.r.t. the optical axis.

Both alternatives result in the same measurements and the same sensing matrix **A**, yet correspond to different optical (or analog) acquisition methods.

9.2 An Out-of-Focus Random Convolution Scheme for Multispectral Compressive Imaging

The Panchromatic Case

Given the sensing operator in (9.2) we now proceed to implementing it in an optical scheme as in [32]. Consider the acquisition of a panchromatic scene $\mathbf{x} \in \mathbb{R}^{n_x \times n_y}$ (*i.e.*, $n = n_x n_y$). We aim at acquiring $m = m_x m_y$ measurements by random convolution of the scene, and assume that these correspond to $m_x \times m_y$ pixels taken from a single snapshot with a low-resolution sensor, *i.e.*, the *Focal Plane Array (FPA)*. From (9.2) we assume a coded aperture **M** of size $s_x = n_x + m_x - 1, s_y = n_y + m_y - 1$ so that the central part of the linear convolution $\mathbf{M} * \mathbf{x}$ will correspond to $m_x \times m_y$ measurements obtained with **x** fully lapped over the coded aperture.

An imaging architecture that implements this operation is depicted in Fig. 9.2: from left to right, an image of the scene x is formed on an aperture plane, in front of the coded aperture. The size of this image must match the spatial resolution (*i.e.*, the size of the pixels) of the coded aperture. This image irradiates toward the coded aperture M, that is placed out-of-focus so that each $n_x \times n_y$ sub-pattern of coded aperture elements is illuminated by a replica of x. Finally, a lens focusing at ∞ is set after the coded aperture, so that every ray that hits and passes it at a certain angle θ w.r.t. the optical axis is focused on the same point at the focal plane. Neglecting the impact of diffraction at the aperture, in this geometrical model each sensor pixel in the FPA sees the sum of the intensity values of a scene $x(u,v) \approx$ x modulated by an $n_x \times n_y$ submatrix of M, *i.e.*, it measures one element of the convolution M $*$ x.

To finalise the implementation of the sensing operator, a simple choice is represented by taking $\mathbf{P}^\Omega$ in (9.2) as a selection matrix so that Ω corresponds to the central elements of M $*$ x, *i.e.*, those that overlap with the $m_x \times m_y$ pixels of the FPA. This choice of Ω strongly increases the correlation between measurements; it is worth noting that this correlation plays a fundamental role in reducing the quality of the recovered MS cube, and as a general guideline strategies should be adopted to minimise it. This observation is also noted by the authors of [32], where it is argued that either a 25% fill-factor sensor or pixel binning (*i.e.*, averaging the outputs of multiple sensor pixels) should be used.

In addition, this correlation will only be made worse by the PSF of the focusing lens and the convolution kernel caused by diffraction, that will inevitably arise as a result of placing the coded aperture out-of-focus in the optical scheme. While a physical optics model of these effects may be conceived, we will opt for a simpler, non-parametric estimate of the PSF seen at the FPA.

A numerical evaluation of this scheme is provided in [32]; our emphasis is on the evaluation and development of our MS version of this optical scheme, which follows in the next Section.

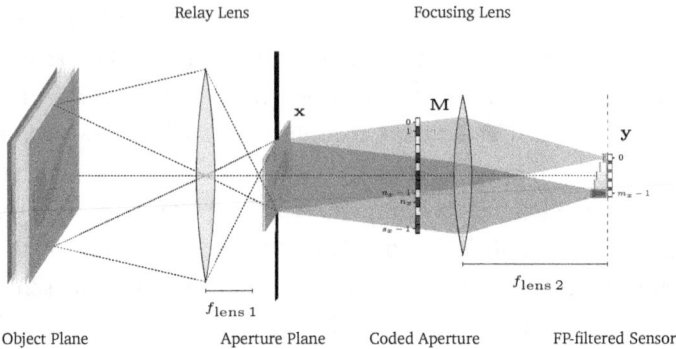

Figure 9.3: Optical scheme (sagittal section) of a multispectral compressive imager by out-of-focus random convolution. The rays traced before the coded aperture are highlighted with the colour of the corresponding FP-filtered sensor pixel.

The Multispectral Case

The extension of the previous scheme to the MS case, *i.e.*, to the acquisition of a data volume $\mathbf{x} \in \mathbb{R}^{n_x \times n_y \times n_\lambda}$ is obtained by placing a MS snapshot sensor such as that described in [195] at the focal plane of the previous scheme. Clearly, the n_λ spectral bands on which the FP-filters deposited on the sensor are tuned must be aligned with those that are assumed of interest in the sensing operation. By doing so, each pixel on the FPA sees a different slice of the cube (see Section 1.1.3) modulated by the coded aperture, *i.e.*, the sensing operator of (9.2) is extended by two considerations:

1. since a total of m measurements is required from the FPA, the latter would produce m/n_λ measurements per band in a single snapshot acquisition. Rather than doing so, we here consider the possibility of partitioning the acquisition of $\mathbf{y} \in \mathbb{R}^m$ by taking multiple snapshots with m_s different aperture patterns, *i.e.*, $\{\mathbf{M}_{(t)}\}_{t=0}^{m_s-1}$. This implies the use of a *programmable* coded aperture. For each band, only $m/n_\lambda m_s$ measurements are therefore required at each snapshot. In addition, taking multiple snapshots with different aperture patterns will reduce

the measurements' correlation;

2. different arrangements of the FP-filters will affect the correlation properties of y. We could either (i) assume a *tiled* layout, i.e., a set of square FP-filtered tiles on the FPA mapped to sensing a single wavelength; (ii) opt for a *mosaic* layout, i.e., the iteration of a pattern of different FP-filters across the whole sensor area. Both FP-filter layout options are simply modelled by a set of selection matrices $\mathbf{P}^{\Omega_l}$, with Ω_l the sensor pixels corresponding to the l-th FP-filter wavelength. In addition, the mosaic layout intrinsically decorrelates measurements of the same band by forcing a spatial gap between them; it is therefore expected to achieve better performances in the implementation of a random convolution scheme.

The acquisition of the data volume is again described as $\mathbf{y} = \mathcal{A}(\mathbf{x})$ with $\mathbf{y} \in \mathbb{R}^m, m = m_x m_y m_s$ being the measurements sampled by the sensor in all m_s snapshots. These measurements are partitioned by the FP-filters in n_λ wavelengths, with $\mathcal{A} : \mathbb{R}^{n_x \times n_y \times n_\lambda} \to \mathbb{R}^m$ describing this mapping. This can be obtained from (9.2) by letting $\mathbf{y}_{(l,t)}$ denote the measurements related to the l-th wavelength and t-th snapshot, i.e.,

$$\mathbf{y}_{(l,t)} = \mathbf{P}^{\Omega_l} \text{vec}\left(\mathbf{M}_{(t)} * \mathbf{x}\right), \; \mathbf{y}_{(l,t)} \in \mathbb{R}^{\frac{m_x m_y}{n_\lambda}} \quad (9.5)$$

where the 2D linear convolution $*$ is applied separately and identically to each slice of $\mathbf{x}$. Thus, $\mathbf{y}$ is obtained by collecting and ordering all the $\mathbf{y}_{(l,t)}$ for $l = \{0, \ldots, n_\lambda - 1\}, t = \{0, \ldots, m_s - 1\}$. This results in

$$\mathbf{y} = \mathcal{A}(\mathbf{x}) = \begin{bmatrix} \mathbf{P}^{\Omega_0} \text{vec}\left(\mathbf{M}_{(0)} * \mathbf{x}\right) \\ \vdots \\ \mathbf{P}^{\Omega_{n_\lambda - 1}} \text{vec}\left(\mathbf{M}_{(0)} * \mathbf{x}\right) \\ \mathbf{P}^{\Omega_0} \text{vec}\left(\mathbf{M}_{(1)} * \mathbf{x}\right) \\ \vdots \\ \mathbf{P}^{\Omega_{n_\lambda - 1}} \text{vec}\left(\mathbf{M}_{(m_s - 1)} * \mathbf{x}\right) \end{bmatrix} \quad (9.6)$$

which fully describes the optical processing performed by the scheme in Fig. 9.3 when m_s snapshots of a MS image $\mathbf{x}$ are captured.

Performance Evaluation by Simulation

The scheme of Fig. 9.3 is here simulated for an ideal, tiled FP-filtered sensor with $m_x = m_y = 256$ partitioned in $n_\lambda = 64$ different FP filters. Thus, each snapshot collects $m_x m_y / n_\lambda = 1024$ new measurements per wavelength (*i.e.*, in 32×32 pixel tiles). To test the capabilities of (9.6) as a sensing operator we consider two synthetic MS images of $n_x \times n_y \times n_\lambda = 256 \times 256 \times 64$ voxel with different prior models that will be specified below. By varying the number of snapshots $m_s = \{1, 4, 9, \ldots, 36\}$ we therefore explore the undersampling rates of $m/n = \{1/64, \ldots, 36/64\}$ w.r.t. $n = 2^{22}$ voxel.

As usual, the simulation procedure entails applying $\mathbf{y} = \mathcal{A}(\mathbf{x})$ with $\mathcal{A}$ as in (9.6) and estimating $\hat{\mathbf{x}}$ (or its transform-domain representation $\hat{\mathbf{s}}$) with a suitable convex optimisation algorithm, that in this case depends on the chosen signal model. In more detail, we assumed:

1. a "Mondrian" data volume $\mathbf{x}$ comprised of eight randomly generated cubes with piecewise-constant values, that exhibits by definition a sparse model w.r.t. TV as defined in (1.6). To provide an accurate recovery of the data volume, we considered the general form of Problem 1.8 with a penalty $\mathcal{P}_0(\boldsymbol{\xi}) = \|\boldsymbol{\xi}\|_{\mathrm{TV}}$, along with $\mathcal{P}_1(\boldsymbol{\xi}) = \|\mathbf{y} - \mathcal{A}(\boldsymbol{\xi})\|_2^2$ in (1.27). To handle the solution of this specific problem, we used the TwIST solver [207] with the weights $\gamma_0 = 1$ and γ_1 in (1.27) hand-tuned to yield optimal reconstruction performances;

2. a "Smiley" data volume $\mathbf{x}$ comprised of a grayscale cartoon image modulated in the wavelength domain by four randomly-generated spectral profiles; this implies a low-rank signal model with rank $\varrho = 4$. In addition to the low-rank prior on the data volume $\mathbf{x}$ when suitably rearranged in slices, we applied (see, *e.g.*, [208]) a joint-sparse signal model by decomposing the MS image slices separately in the spatial domain with $\mathbf{D} = \mathbf{D}_{x,y} \otimes \mathbf{I}_{n_\lambda}$, $\mathbf{D}_{x,y} = \mathbf{D}_x \otimes \mathbf{D}_y$ a 2D Haar orthonormal DWT. Thus, for the l-th slice $\mathbf{x}_l$ we have $\mathbf{s}_l = \mathbf{D}_{x,y}^* \mathrm{vec}(\mathbf{x}_l)$ and by Definition 1.7 we

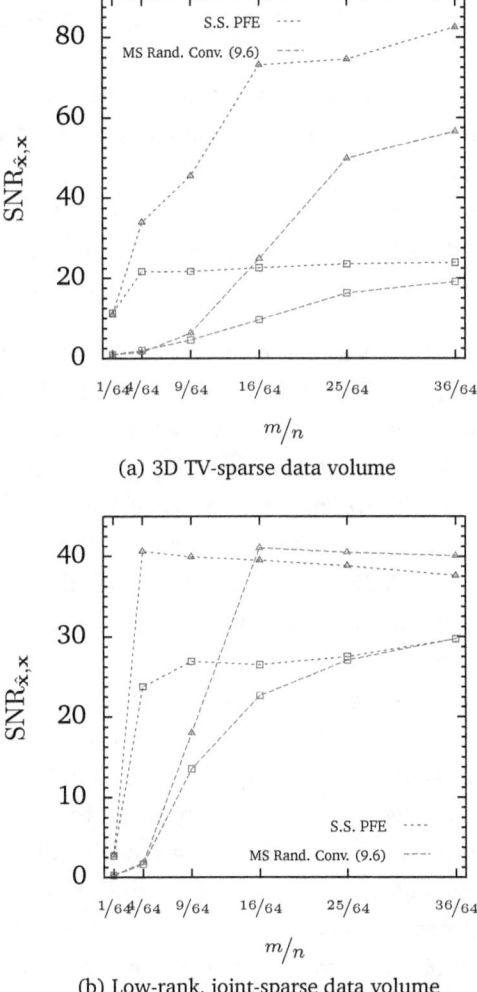

(a) 3D TV-sparse data volume

(b) Low-rank, joint-sparse data volume

Figure 9.4: Simulated recovery performances of a multispectral compressive imager by out-of-focus random convolution *versus* the spread-spectrum PFE; AWGN is added to attain a measurement SNR of $20\,\text{dB}$ (squares) and $80\,\text{dB}$ (triangles).

obtain $\mathbf{S} = \{\mathbf{s}_l\}_{l=0}^{n_\lambda-1}$ with joint-sparsity level $k = \|\mathbf{S}\|_{2,0} = 4096$ (*i.e.*, $k/n = 1/1024$).

As for the recovery algorithm, we plug into Problem 1.8 the penalties $\mathcal{P}_0(\Xi) = \|\mathbf{D}_{x,y}^*\Xi\|_{2,1}$ for joint-sparsity, and the nuclear norm $\mathcal{P}_1(\Xi) = \|\Xi\|_*$ to promote a low-rank model. The formulation of this recovery problem is completed with the data fidelity constraint $\mathcal{C}(\boldsymbol{\xi}) = \{\boldsymbol{\xi} \in \mathbb{R}^{n_x \times n_y \times n_\lambda} : \|\mathbf{y} - \mathcal{A}(\boldsymbol{\xi})\|_2^2 \leq \varepsilon^2\}$, where $\mathcal{A}$ is the sensing operator described by (9.6). The weights were chosen as $\gamma_0 = 1, \gamma_1 = \varrho$, and the solver of reference for this case was the algorithm by Chambolle and Pock [209].

For a fair comparison given these priors, we compared the sensing operator in (9.6) against the spread-spectrum PFE [89] of the same size and undersampling rate (*i.e.*, as in (1.16)) as applied separately to each slice in the MS cube; this RME is here adopted as a computationally lightweight alternative to the ideal RGE of the same dimensions. Finally, two noise levels were tested to investigate the robustness of this sensing operator to AWGN when recovered with the same methods, since an analog-domain implementation of (9.6) will be affected by a non-negligible amount of noise.

The results are reported in terms of $\mathrm{SNR}_{\hat{\mathbf{x}},\mathbf{x}}$ as a function of the undersampling rate in Fig. 9.4b, 9.4a. There, it can be seen that higher $\mathrm{SNR}_{\hat{\mathbf{x}},\mathbf{x}}$ values are achieved by the spread-spectrum PFE in all cases as m increases; this recovery quality quickly saturates at the maximum $\mathrm{SNR}_{\hat{\mathbf{x}},\mathbf{x}}$ level imposed by the presence of AWGN. When comparing the SNR performances as a function of m, the same quality level is achieved for significantly larger values in the case of (9.6). This only confirms that the random convolution operator is a sub-optimal choice in the RIP sense. In addition, there is a strong variability in the $\mathrm{SNR}_{\hat{\mathbf{x}},\mathbf{x}}$ between the two signal models adopted for this evaluation; thus, the required number of snapshots will depend on the complexity of the data volume being acquired.

Even with the anticipated quality limits of Fig. 9.4b, 9.4a we favour the simplicity of the layout in Fig. 9.3 and proceed to its implementation as a proof-of-concept of MS imaging by CS.

9.3 Design and Implementation of a Prototype for Compressive Imaging by Out-of-Focus Random Convolution

9.3.1 The Optical System

The implementation details of this imager lie outside of the signal processing perspective of this thesis, and are therefore only summarised here. The main step in its optical-level design is the selection of a suitable *Spatial Light Modulator (SLM)*, *i.e.*, a programmable coded aperture whose technology, resolution and size are critical and must be carefully matched with the FPA. Digital micro-mirror SLMs such as those used in the implementation of the single-pixel camera [20] are not suitable for a realisation of out-of-focus random convolution, as they introduce some angular uncertainty due to micro-mirror tilting. This limit makes such a technology incompatible with the principle of Fig. 9.2,9.3 where different angles encode different convolution elements. Thus, we chose a liquid-crystal-on-silicon SLM for its high light transmittance performances [210] and the capability of mounting and aligning it with suitable precision.

By matching the chosen SLM with the available FP-filtered sensor arrays, an optical system providing a random convolution was assembled as in Fig. 9.5a and 9.5b. Note that by replacing the FP-filtered FPA with a standard one we seamlessly obtain a panchromatic imager that substantially realises with a programmable SLM the scheme in [32].

Since the SLM operates on polarised light, a polarising beam-splitter (see [211]) must be introduced and forces a layout in which the scene is reflected toward the SLM (*i.e.*, a reflective layout); with this, the optical design assumes an "L" configuration, as can be seen in the implemented scheme. As a result, the alignment of the optical elements is slightly more complex and requires considering that a residual mismatch will always exist between the ideal operator (either (9.2) or (9.6)) and its actual optical-level effect. In the following, this mismatch is modelled by the estimation of a PSF.

We note that up to this point we have omitted any resolution specification, as the assembled prototype is intended as a platform for testing a variety of resolution configurations in a flexible manner, *i.e.*, the total area of the SLM and FPA exceeds the active area used in the evaluation. The resolution parameters will be specified in Section 9.4.

9.3.2 Diffraction Kernel and Point Spread Function Estimation

As well anticipated by [32], the main optical-level limitation of this scheme is the impact of diffraction that inevitably occurs at the coded aperture, which is essentially an array of random square apertures on a uniform grid. By Fourier optics [203, Chapter 4] when a single square aperture of small width is illuminated by a spatially and temporally coherent plane wave, a Fraunhofer (*i.e.*, far-field) diffraction pattern is formed at the focal plane of a lens focusing objects at infinity (*i.e.*, as the one placed in front of the FPA in Fig. 9.2, 9.3). Thus, the optical system must account for the presence of an additional 2D convolution kernel that describes the diffraction pattern produced by a single element of the coded aperture on the focal plane when illuminated with *spatially incoherent* light; in other words, the effect of diffraction at a single aperture element is here modelled as an optical filter (see [203]), as partially explained in [32].

However, in addition to this expected effect the PSF of the lens should also be taken into account, as well as possible focusing errors and the wavelength dependency of both the lens' PSF and the above diffraction kernel. The composition of these three effects is not trivially modelled; thus, we here assume a system-level perspective and aim at estimating (under a linear shift-invariant hypothesis) a single PSF $\hat{\mathbf{h}} \in \mathbb{R}^{n_h \times n_h}$ discretised at the spatial resolution of the FPA and modelling all the aforementioned non-idealities. This is done by means of a panchromatic FPA, with which a more realistic panchromatic-case sensing operator is

$$\mathbf{y} = \mathbf{P}^\Omega \text{vec}\left(\hat{\mathbf{h}} * [\mathbf{M} * \mathbf{x}]\right) + \boldsymbol{\nu}, \ \mathbf{y}, \boldsymbol{\nu} \in \mathbb{R}^m \qquad (9.7)$$

Design and Implementation

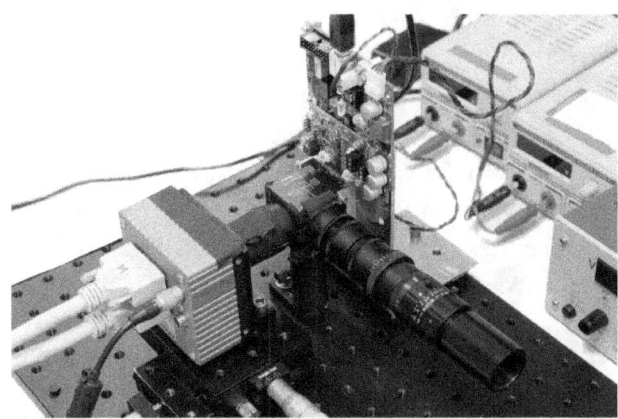

(a) Perspective view

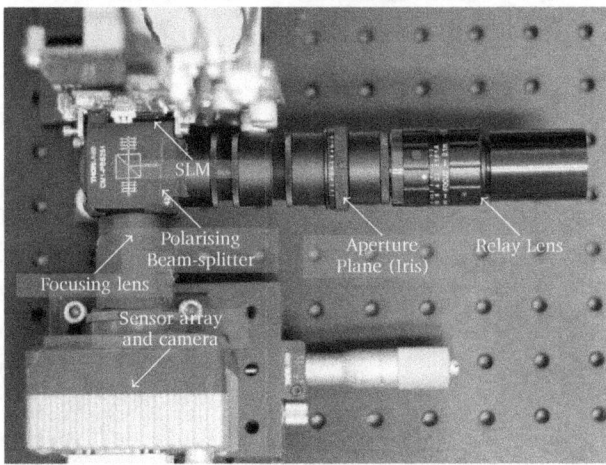

(b) Top view

Figure 9.5: A picture of the designed imaging system on the optical table; the figure also depicts the SLM control board and the camera on which the IMEC FP-filtered sensor array is mounted.

Two procedures may then be applied to estimate $\hat{\mathbf{h}}$:

1. the first procedure entails injecting into the optical system some known, printed or displayed test patterns so that their image is correctly focused at the aperture plane, and measuring their response at the FPA with a full-size Nyquist-rate sampling. Thus, given a known test pattern $\mathbf{x} \in \mathbb{R}^{n_x \times n_y}$ and the ideal coded aperture pattern $\mathbf{M} \in \{-1, +1\}^{s_x \times s_y}$ programmed[7] on the SLM we measure $\mathbf{y}$ as in (9.7) with $m = m_x m_y = n$, i.e., at Nyquist rate. Once these measurements are collected we substantially have to solve a linear inverse problem with $\hat{\mathbf{h}}$ as the unknown;

2. as an alternative, we could resort to taking the same amount of Nyquist-rate measurements with a URA pattern at the SLM, as mentioned in Section 9.1.1. This would yield an estimate $\hat{\mathbf{x}} \approx \hat{\mathbf{h}} * \mathbf{x}$ by deconvolving the measurements. Moreover, when the SLM is illuminated by $x(u, v) \approx \delta(u, v)$, i.e., by a point source such as an optical fibre placed at the aperture, this will yield $\hat{\mathbf{x}} \approx \hat{\mathbf{h}}$.

Thus, both methods substantially yield a set of Nyquist-rate measurements $\mathbf{y} \in \mathbb{R}^n$ that verify (9.7) with $\hat{\mathbf{h}} \in \mathbb{R}^{n_r \times n_r}$ as the only unknown; we may then use sparsity to promote a rapid DFT magnitude decay of the solution. Given its nature the kernel $\hat{\mathbf{h}}$ is indeed expected to have a low-pass profile w.r.t. the 2D DFT. Thus, PSF estimation is rephrased as

$$\hat{\mathbf{h}} = \operatorname*{argmin}_{\boldsymbol{\xi} \in \mathbb{R}^{n_h \times n_h}} \|(\mathbf{F}_{n_h} \otimes \mathbf{F}_{n_h}) \mathrm{vec}\,(\boldsymbol{\xi})\|_1 + \gamma \, \|\mathbf{y} - \mathrm{vec}\,([\mathbf{M} * \mathbf{x}] * \boldsymbol{\xi})\|_2^2 \tag{9.8}$$

which is a standard convex optimisation problem; when (9.8) is plugged into the general (1.27) and solved with, e.g., Douglas-Rachford splitting [113] we obtain the desired estimate of $\hat{\mathbf{h}}$.

This PSF estimation procedure was applied by using a URA pattern (yet observing that a similar PSF was produced by the other estimation

---

[7] As mentioned, this actually involves capturing two frames per pattern as in (9.4).

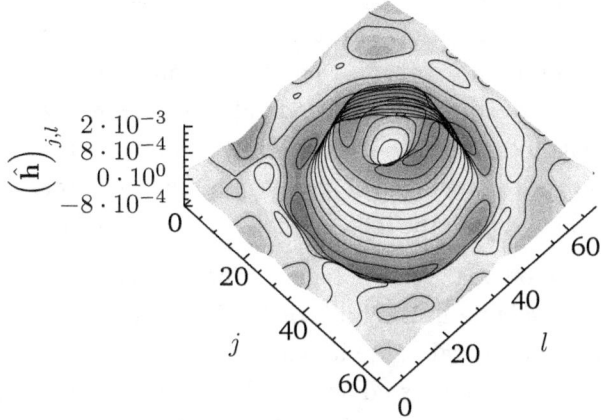

(a) Estimated PSF; equivalent FPA area: 70×70 pixel.

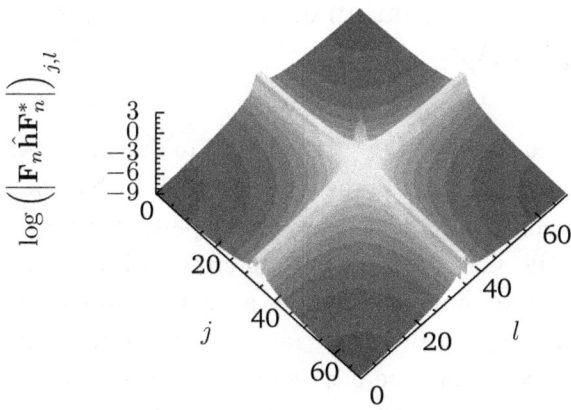

(b) 2D DFT magnitude of the estimated PSF (logarithmic scale); a strongly low-pass behaviour can be observed.

Figure 9.6: PSF estimation for the out-of-focus multispectral compressive imager.

procedure) in the system configuration of Fig. 9.5b, yielding $\hat{h}$ as reported in Fig. 9.6a and refined by (9.8) (here $n_h = 70$).

From the 2D DFT of this PSF we observe a substantially low-pass response, indicating that the effect of the latter is substantially that of limiting the bandwidth of the sensing operator $\mathcal{A}$; thus, the effective random convolution operator in (9.7) modulates the scene at a rate smaller than the Nyquist rate of x. This will cause a further decrease in the performances of random convolution, as the measurements will be even more correlated than the ideal models (9.2) and (9.6).

With this information on the PSF at hand, a closer-to-reality sensing operator model is available in the form of (9.7); it is then possible to proceed to an experimental phase in which the actual, achievable capabilities of the imager are tested against the impact of noise and diffraction. Other mismatches, such the calibration of different gains at each sensor pixel, are left for future improvements.

9.4 Performance Evaluation of Panchromatic and Multispectral Imaging

The experiments illustrated in this Chapter refer to the measurements obtained by two configurations of the scheme in Fig. 9.5b, depending on the type of sensor used as a FPA (either panchromatic or FP-filtered). Since the non-idealities encountered during calibration become critical in the implementation of a high-resolution scheme for either panchromatic or multispectral compressive imaging, the spatial resolution documented here is significantly lower than the one anticipated in the simulations of Section 9.2. In addition, the maximum number of wavelengths considered by the FP-filtered sensor used in these experiments is $n_\lambda = 16$. The most accurate results obtained so far are therefore documented, leaving a large room for improvement and some notes on how it may be obtained by refining the sensing operator models to match the reality of their physical implementation.

9.4.1 Panchromatic Compressive Imager

In the case of a panchromatic imager using the hardware architecture in Fig. 9.5b we target a scene resolution[8] of $n_x \times n_y = 128 \times 128$ pixel. With this scene resolution, we consider collecting a variable amount of measurements at the FPA given the programmable nature of the prototype; this implies a flexible reconfiguration of the coded aperture pattern programmed at the SLM, of dimensions $s_x \times s_y = n_x + m_x - 1 \times n_y + m_y - 1$. This allows us to acquire an increasingly large fraction of the $n = n_x n_y = 16384$ Nyquist-rate values in $\mathbf{x} \in \mathbb{R}^{n_x \times n_y}$ as in (9.7).

The scenes evaluated here are comprised of objects illuminated with a halogen light bulb, irradiating high-intensity white light on each test image; for each case we collect $\mathbf{y}^+$ and $\mathbf{y}^-$ with the same policy of (9.4), yet accounting for the estimated PSF as in (9.7); this yields a measurement vector $\mathbf{y}$. As test images we use (i) a printed white "λ" on a black background and (ii) a paper cup. To perform signal recovery of a scene $\hat{\mathbf{x}}$ we solve aBPDN in (1.24); to reduce the impact of noise [139] and promote sparsity in an overcomplete dictionary, we let $(\mathbf{D}, \mathbf{D}^\star)$ be the synthesis and analysis operators of the 2D Daubechies-4 UDWT with $J = 9$ sub-bands, $p = (J+1)n$. The inequality constraint of BPDN is hand-tuned to yield the best visual quality, and corresponds to a measurement noise norm $\|\boldsymbol{\nu}\|_2 \leq \varepsilon$ where $\varepsilon - 25\,\text{dB}$. The chosen solver for Problem 1.6, in which this configuration is plugged, is the algorithm of Chambolle and Pock [209].

Regrettably, evaluating the $\text{SNR}_{\hat{\mathbf{x}},\mathbf{x}}$ is not reliable, as an exact ground truth for $\mathbf{x}$ is not available in the exact position of the test targets (and even a small shift or rotation of the actual scene w.r.t. the assumed ground truth could lead to small SNRs). A visual assessment of the recovery performances can however be given from the results in Fig. 9.7. From these, it can be seen that the recovery quality does not improve after capturing 31.25% of the Nyquist-rate measurements, *i.e.*, it saturates to a scene showing some residual error in the form

---

[8] To reduce the impact of diffraction, each SLM pixel is actually a block of 4×4 physical pixels; the FPA readings are similarly obtained by binning a block of 4×4 physical pixels. This does not change the compression rate of the scheme.

of blurred, low-pass noise; this effect is compatible with residual noise and mismatches in the sensing operator, as increasing the measurements' dimensionality does not significantly mitigate its impact (as it would for non-sparse signals). In order to improve upon these results, a more accurate calibration procedure will be required to refine the sensing operator $\mathcal{A}$; such a one-time calibration could, *e.g.*, rely on another SLM injecting a large number of test patterns directly at the aperture plane, in a fashion similar to [212]. This will eventually allow a higher scene resolution by correcting the ideal sensing operator (9.2) into a model that is more accurate than (9.7).

9.4.2 Multispectral Compressive Imager

In the MS case, the panchromatic sensor is replaced by a FP-filtered one as manufactured by IMEC in the camera of Fig. 9.5b. The FP filter pattern is a mosaic of 4×4 pixel, each sensing a different wavelength. This elementary pattern is here defined as a *macropixel*. At each snapshot we sample $m_x \times m_y = 64 \times 64$ macropixel $= 256 \times 256$ pixel partitioned in $n_\lambda = 16$ wavelengths.

Thus, we target the recovery of a scene of $n_x \times n_y \times n_\lambda = 128 \times 128 \times 16$ voxel with 16 VIS bands, as allowed by the chosen sensor[9]. This will require coded apertures of size $s_x \times s_y = 383 \times 383$ pixel and a maximum of $m_s < 4$ snapshots (also corresponding to a maximum of four programmable coded aperture patterns), with $m_s = 4$ corresponding to a Nyquist-rate sampling of the MS cube.

Clearly, this system can be configured to have larger coded apertures, larger FPAs and less snapshots, or conversely smaller coded apertures, smaller FPAs and more snapshots, in a completely flexible manner; it is worth noting that actually, partitioning the acquisition between many snapshots reduces the measurements' correlation, and is therefore expected to have a beneficial effect on the recovery quality.

However, the application of FP filters reduces the amount of photons received at each pixel during the exposure; thus, the general

---

[9] Two bands were actually excluded in the sensing, due to the redundancy of two FP filters according to the sensor's specifications.

Performance Evaluation

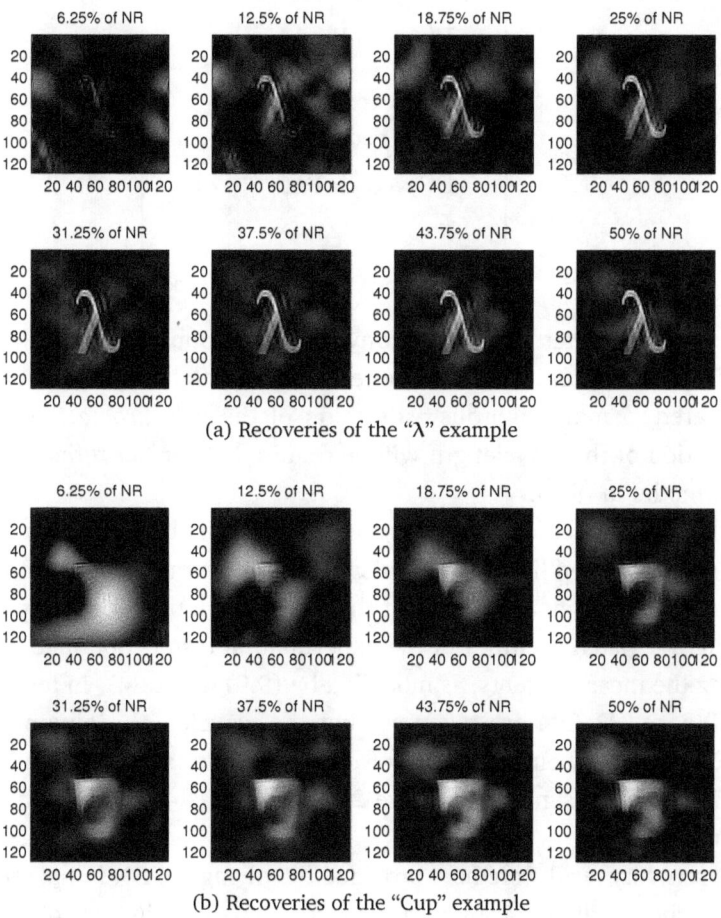

Figure 9.7: Recovered panchromatic images as a function of the undersampling w.r.t. the Nyquist rate; the estimated PSF is used in the sensing model to improve the recovery quality.

impact of measurement noise is expected to be higher, at the expense of a lower recovery quality. In addition, the estimated PSF $\hat{\mathbf{h}}$ is practically downsampled as an effect of the spatial distance between pixels of the same wavelength; thus, the sensing operator becomes

$$\mathbf{y} = \mathcal{A}(\mathbf{x}) = \begin{bmatrix} \mathbf{P}^{\Omega_0} \text{vec}\left(\hat{\mathbf{h}} * \mathbf{M}_{(0)} * \mathbf{x}\right) \\ \vdots \\ \mathbf{P}^{\Omega_{n_\lambda - 1}} \text{vec}\left(\hat{\mathbf{h}} * \mathbf{M}_{(0)} * \mathbf{x}\right) \\ \mathbf{P}^{\Omega_0} \text{vec}\left(\hat{\mathbf{h}} * \mathbf{M}_{(1)} * \mathbf{x}\right) \\ \vdots \\ \mathbf{P}^{\Omega_{n_\lambda - 1}} \text{vec}\left(\hat{\mathbf{h}} * \mathbf{M}_{(m_s - 1)} * \mathbf{x}\right) \end{bmatrix} \qquad (9.9)$$

where $*$ still denotes 2D linear convolution as applied separately to each of the slices of $\mathbf{x}$. We note that, although slightly more complicated, a more rigorous estimation of the *polychromatic* PSF as a function of the wavelength will be required to further refine the sensing model in (9.9).

In this case, the chosen example scene is a colour chart illuminated with a white halogen light bulb. The ground truth corresponding to it is not available for an evaluation of the $\text{SNR}_{\hat{\mathbf{x}}, \mathbf{x}}$; we therefore simply illustrate the visual accuracy of the experimental results obtained by plugging the measurements, as modelled by (9.9), in (1.24). In terms of dictionary, $(\mathbf{D}, \mathbf{D}^*)$ are taken so that the wavelength domain is analysed with $\mathbf{D}_\lambda$ being the DCT of dimensionality $n_\lambda = 16$, while $(\mathbf{D}_{xy}, \mathbf{D}^*_{xy})$ are chosen as the above 2D Daubechies-4 UDWT with 9 sub-bands.

The recovery results are here provided by solving aBPDN initialised with the chosen dictionary and a data fidelity constraint set to $\varepsilon = -13\,\text{dB}$; its solution with the Chambolle and Pock algorithm [209] yields the MS images depicted in Fig. 9.8 for $m_s = 1$ (25% of Nyquist rate), Fig. 9.9 for $m_s = 2$ (50% of Nyquist rate) snapshots. There, the squares depicted in the colour chart can be appreciated. It must however be noted that only a slight variability in the slices is perceived; this is mainly due to the fact that light leakage from wavelengths outside the VIS range has not been suitably removed by a global VIS

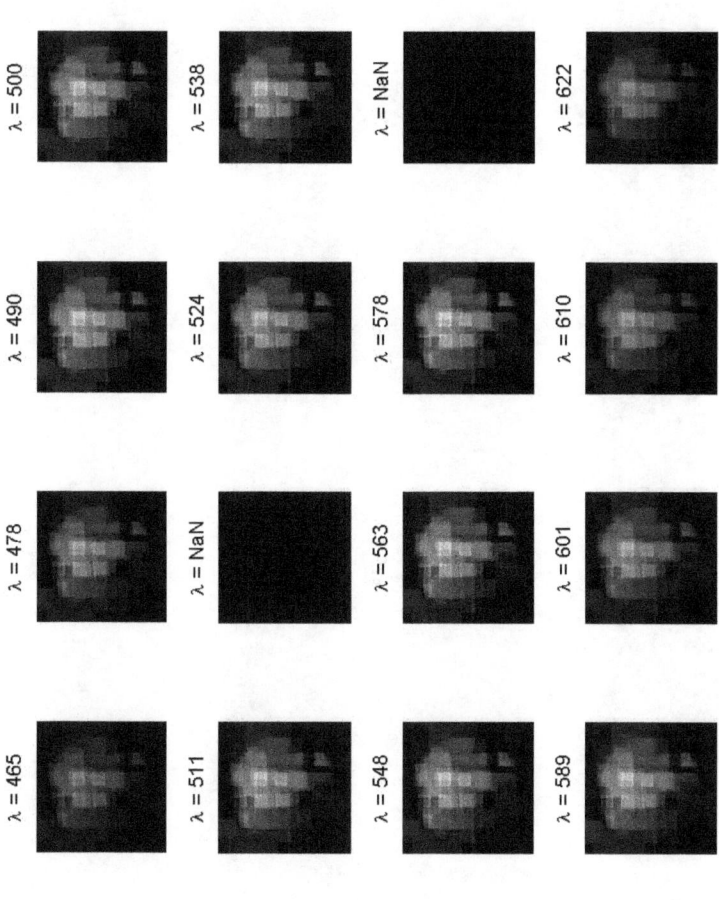

Figure 9.8: Recovered MS slices of $\hat{x}$ using 25% of the Nyquist-rate measurements; the slices' colour map is adapted to the centre wavelength of each FP filter; the missing wavelengths are reported as $\lambda = $ NaN.

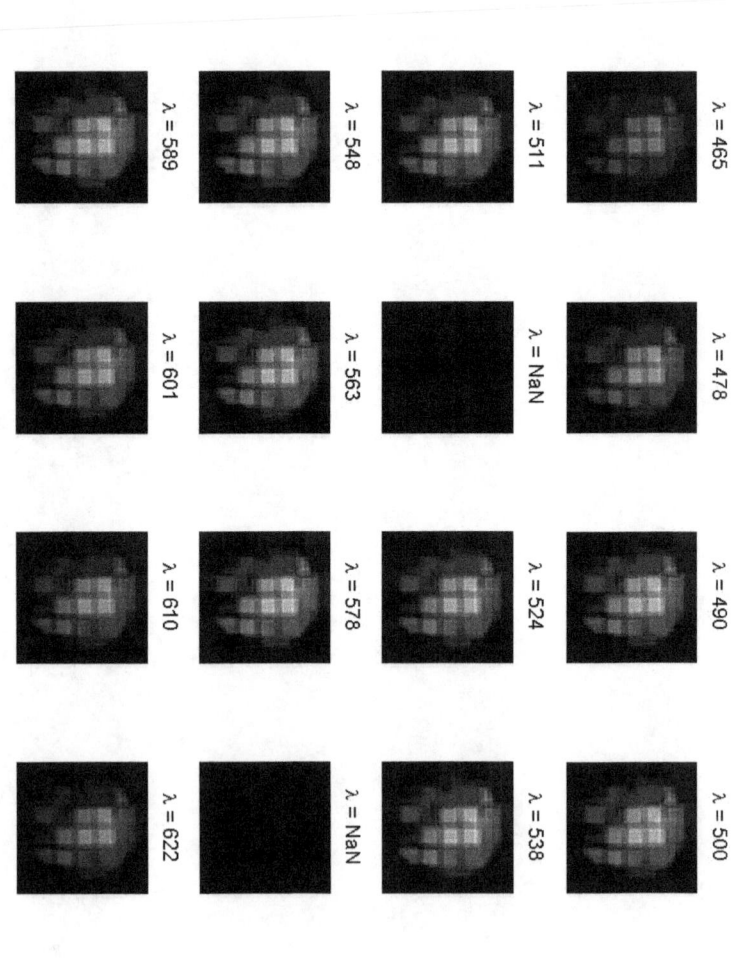

Figure 9.9: Recovered slices of $\hat{\mathbf{x}}$ using 50% of the Nyquist-rate measurements; the slices' colour map is adapted to the centre wavelength of each FP filter; the missing wavelengths are reported as $\lambda = \text{NaN}$.

band-pass cut-off filter in these experiments. This causes the observed visual similarity between the slices, although they differ both in terms of total light intensity per wavelength, as well as in the spatial domain. Thus, a MS sensing capability is provided by the scheme in Fig. 9.5b when using a FP-filtered FPA.

The recoveries of Fig. 9.8, 9.9 display some sharp borders in the colour chart squares; thus, we may conclude that the devised MS imager is capable of operating in an undersampling regime, although the observed recovery quality is indeed limited by the mentioned non-idealities. An improvement w.r.t. the present quality will be possible by addressing them with a more accurate calibration procedure, as well as the addition of a suitable cut-off filter in front of the relay lens in Fig. 9.5b to emphasise the difference between the recovered MS slices.

Summary

- CS may be performed by random convolution, a sensing operator that is particularly convenient to implement in an optical scheme.

- An optical scheme for MS imaging by random convolution was built by following the initial observations of [32]. The scheme involves the use of a FPA that is filtered by a layer of Fabry-Pérot cavities. A numerical simulation of the scheme shows that an undersampling regime is possible even when CS is not applied in the wavelength domain.

- A prototype implementing random convolution by means of a SLM was assembled and evaluated. Some alignment and calibration procedures were introduced to verify that the optical setup implements one-to-one the expected sensing operator. A particular emphasis was dedicated to the impact of diffraction, modelled as a convolution that was jointly estimated with other non-idealities in a non-parametric PSF.

- A performance assessment of this prototype was carried out with a panchromatic sensor and using simple test scenes. The experimental results showed limited recovery quality, where it is seen that the losses of due to undersampling are dominated by those due to measurement noise and sensing operator mismatches, *i.e.*, setting the undersampling rate up to the Nyquist rate of the image does not improve the recovery.

▶ A similar assessment was carried out with a FP-filtered FPA; the recovery of a MS image is indeed possible with higher visual recovery quality due to the redundancy of the data volume in the wavelength domain. However, little variability is seen due to light leakage from outside the pass-band of the FP filters. Nevertheless, this variability is present and a suitable calibration against a white reference and the addition of spectral cut-off filters will yield full MS sensing capability.

▶ While there are clear limits in these early results, a large margin of improvement remains to reach stable signal recovery performances. In fact, it is in the matching between the optical effect of the setup and the mathematical model, which was only outlined here, that the main issues and discrepancies arise: the investigation of new calibration procedures for a refinement of the sensing operator is therefore crucial to attain near-ideal performances after signal recovery.

Conclusions

In this thesis we have presented a number of extensions and methods that elaborate the concept of Compressed Sensing into a variety of improvements for prospective applications; in particular, we have focused on:

- **Maximum Energy Sensing Matrix Designs**: we have proposed a method for the design of random sensing matrix ensembles based on a proxy for information extraction, namely, a maximisation of the measurements' energy. The *a priori* hypothesis made on the signal ensemble being acquired is that it is *localised*, *i.e.*, unevenly distributed in its domain so that by a simple analysis of its correlation matrix an improvement can be obtained in adapting the statistical distribution of the row vectors of a random sensing matrix.

    Thus, novel optimisation problems were presented as possible extensions to this energy maximisation problem, first addressed in [27, 29]. The non-trivial matter of showing the optimality of this choice was numerically explored by means of the Donoho-Tanner empirical phase transition w.r.t. Basis Pursuit and against sparse and localised random vectors on which the adaptation was carried out. Intuition on how this design is advantageous only when the signal ensemble complies with the hypothesis was provided.

The experimental results highlight a significant improvement in the capability of recovering sparse and localised vectors from statistically designed Random Matrix Ensembles with adapted second-moments. This choice allows to recover signals with less sparse representations if they are sufficiently localised, and *vice versa* of recovering sparse vectors with less measurements. These results are related to [2].

- **Maximum Entropy Sensing Matrix Designs**: another adaptation of sensing matrix designs w.r.t. localised signals was explored, in the particular case of a fixed design space of deterministic sensing vectors that are not necessarily incoherent w.r.t. the atoms of the dictionary in which a signal is analysed. In this case, the measurements' differential entropy was used as a proxy for information extraction. This led to the formulation of an entropy maximisation problem, whose solution is NP-hard in general. To overcome this issue, we have proposed a heuristic algorithm to construct a pool of sensing matrices by selection of a fixed-cardinality subset of sensing vectors.

  The result of this lightweight procedure is an optimised sensing matrix w.r.t. a localised signal ensemble. The experimental results also highlight that the adapted sensing matrix is capable of delivering a significant signal recovery improvement from the same amount of measurements as an Isotropic Random Bernoulli Ensemble. This indicates that an increment of the measurements' entropy augments the amount of information embedded in a set of undersampled measurements. These results are related to [4].

- **Low-Complexity Signal Encoding**: an immediate application of the design flow in [2] is in the possibility of using Compressed Sensing as a digital-to-digital encoding technique, with the non-negligible advantage of using only $\mathcal{O}\left(kn \log n/k\right)$ additions and subtractions when a suitably adapted Anisotropic Random

Bernoulli Ensemble is used to draw sensing (*i.e.*, encoding) matrices. In addition, adopting such a maximum energy encoding matrix design when the correlation matrix of the signal being compressed can be regarded as a stationary property is in a sense analogous to performing a Karhunen-Loève Transform, yet with antipodal-valued projection vectors that allow its implementation by low-complexity multiplierless operations.

The result of applying Compressed Sensing with maximum energy encoding matrices was explored in comparison with Set Partitioning In Hierarchical Trees, a highly efficient algorithm for digital signal compression, as well as traditional, yet optimally-tuned Huffman Coding of the PCM representation of ECG signals. Particular emphasis was deserved to the choice of a quantisation policy for compressive measurements. This led to observing compression rates that approach those of Set Partitioning In Hierarchical Trees, yet are capable of being implemented with a minimum amount of digital computations (*i.e.*, by a single digital accumulator) [3, 9].

▶ **Average Recovery Performances in the Presence of Perturbations**: the matter of predicting the average performances of Compressed Sensing in the presence of matrix perturbations was taken into consideration, as a useful quantification for prospective applications of Compressed Sensing that are subject to disturbances. In this context, an index of the average performances of Basis Pursuit and Basis Pursuit with Denoising was formulated in the hypothesis that a perturbation of relatively small entity is suffered by the entries of the sensing matrix.

The devised index requires a singular value estimation step that is related to the sensing and perturbation matrix; in absence of alternatives, this estimation must be done by Monte Carlo simulation of the involved Random Matrix Ensembles. As an assessment of the configurations in which the obtained performance estimate is reliable, we have provided an extensive dataset for different perturbation matrices and showed how their effect is well anticipated by the proposed average performance index [8].

▶ **Multiclass Encryption by Compressed Sensing**: in the context of providing information security and data protection by means of Compressed Sensing, we have evaluated the properties of an encryption scheme based on the latter when Isotropic Random Bernoulli Ensemble encoding matrices are used. The added value of this approach is in that the same low-complexity observed in signal compression by Compressed Sensing actually grants some security properties directly in the sensing or encoding process. A multiclass extension of this policy allows to distinguish different access modes to the encrypted information content by means of intentional, undetectable and pseudo-random matrix perturbations in the encoding matrices supplied to the receivers.

We have presented bounds on the performances of multiclass encryption, with an emphasis on guaranteeing that lower-class receivers have guaranteed performances falling within a prescribed range, *i.e.*, that their quality is set and controlled by the amount of perturbation introduced in their encoding matrix as a mismatch w.r.t. the true one.

In order to verify that this scheme cannot be statistically cryptanalysed, while in the presence of a non-perfectly secret cryptosystem, we have provided theory and evidence that the ciphertexts, *i.e.*, the compressive measurements with Isotropic Random Bernoulli Ensemble encoding matrices follow an independent and identically distributed Gaussian distribution conditioned on the power (asymptotic case) or energy (non-asymptotic case) of the plaintext, that is the encoded signal. These findings are documented in [5].

On the other hand, while a statistical cryptanalysis yields no useful results to an attacker aside from the energy of the plaintext (that does not *per se* allow a recovery of either the private key or the plaintext) we proposed a theoretical perspective on computational cryptanalysis by counting the number of encoding matrices that match a plaintext-ciphertext pair, *i.e.*, we performed Known-Plaintext Attacks. This class of attacks involved enumerating the solutions of two versions of the subset-sum problem depending on the considered type of malicious user. In fact, we showed (i) how the number of solutions yielded by a Known-Plaintext Attack is very large in the case of an eavesdropper attempting to break a single row vector of the true encoding matrix; (ii) how a smaller, yet still very large amount of solutions is presented to a malicious lower-class user attempting to upgrade its knowledge of the true encoding matrix. This last attack was sided by a signal recovery approach to breaking this matrix, leading to no appreciable quality improvement due to the random nature of the chosen matrix perturbation [6].

▶ **A Multispectral Compressive Imager by Random Convolution:** as part of a collaboration with the Integrated Imagers Division of IMEC, Belgium, the task of designing a multispectral snapshot imager based on Compressed Sensing was undertaken. The chosen optical system architecture was mutuated from a previous panchromatic imager design leveraging the principle of random convolution; yet, the sensing model was extended to account for multispectral Fabry-Pérot-filtered sensors and system-level non-idealities such as the Point Spread Functions of the elements on the optical path.

In this thesis, early experimental results were reported as produced by an optical implementation of this imager, that leaves large room for improvement; future work in this sense will entail a new approach to calibration in order to compensate for mismatches between the mathematical model of the optical system and the actual effect of the optical processing chain on incident light. In this case, a publication of the experimental findings will follow [7].

BIBLIOGRAPHY

[1] V. Cambareri, J. Haboba, F. Pareschi, R. Rovatti, G. Setti, and K.-W. Wong, "A two-class information concealing system based on compressed sensing," in *2013 IEEE International Symposium on Circuits and Systems (ISCAS)*. IEEE, 2013, pp. 1356–1359.

[2] V. Cambareri, M. Mangia, F. Pareschi, R. Rovatti, and G. Setti, "A rakeness-based design flow for analog-to-information conversion by compressive sensing," in *2013 IEEE International Symposium on Circuits and Systems (ISCAS)*. IEEE, 2013, pp. 1360–1363.

[3] V. Cambareri, M. Mangia, R. Rovatti, and G. Setti, "Joint analog-to-information conversion of heterogeneous biosignals," in *2013 IEEE Biomedical Circuits and Systems Conference (BioCAS)*, Oct 2013, pp. 158–161.

[4] V. Cambareri, R. Rovatti, and G. Setti, "Maximum entropy Hadamard sensing of sparse and localized signals," in *2014 IEEE International Conference on Acoustics, Speech and Signal Processing (ICASSP)*, May 2014, pp. 2357–2361.

[5] V. Cambareri, M. Mangia, F. Pareschi, R. Rovatti, and G. Setti, "Low-complexity multiclass encryption by compressed sensing," *IEEE Transactions on Signal Processing*, vol. 63, no. 9, pp. 2183–2195, 2015.

[6] ——, "On known-plaintext attacks to a compressed sensing-based encryption: a quantitative analysis," *IEEE Transactions on Information Forensics and Security*, March 2015, in peer review.

[7] K. Degraux, V. Cambareri, B. Geelen, L. Jacques, G. Lafruit, and G. Setti, "Compressive hyperspectral imaging by out-of-focus modulations and fabry-perot spectral filters," in *iTWIST'14: international International Traveling Workshop on Interactions between Sparse models and Technology*, 2014.

[8] V. Cambareri, M. Mangia, F. Pareschi, R. Rovatti, and G. Setti, "Average recovery performances of non-perfectly informed compressed sensing: with applications to multiclass encryption," in *2015 IEEE International Conference on Acoustics, Speech and Signal Processing (ICASSP)*, April 2015, in press.

[9] V. Cambareri, M. Mangia, R. Rovatti, and G. Setti, "A case study in low-complexity signal encoding: How compressing is compressed sensing?" *IEEE Signal Processing Letters*, March 2015, in peer review.

[10] K. Degraux, V. Cambareri, L. Jacques, B. Geelen, C. Blanch, and G. Lafruit, "Generalized inpainting method for hyperspectral image acquisition," in *2015 IEEE International Conference on Image Processing (ICIP)*. IEEE, 2015, in peer review.

[11] D. L. Donoho, "Compressed sensing," *IEEE Transactions on Information Theory*, vol. 52, no. 4, pp. 1289–1306, 2006.

[12] E. J. Candes and T. Tao, "Decoding by linear programming," *IEEE Transactions on Information Theory*, vol. 51, no. 12, pp. 4203–4215, 2005.

[13] E. J. Candes, J. K. Romberg, and T. Tao, "Stable signal recovery from incomplete and inaccurate measurements," *Communications on pure and applied mathematics*, vol. 59, no. 8, pp. 1207–1223, 2006.

[14] D. L. Donoho and X. Huo, "Uncertainty principles and ideal atomic decomposition," *IEEE Transactions on Information Theory*, vol. 47, no. 7, pp. 2845–2862, 2001.

[15] D. L. Donoho and M. Elad, "Optimally sparse representation in general (nonorthogonal) dictionaries via ℓ_1 minimization," *Proceedings of the National Academy of Sciences*, vol. 100, no. 5, pp. 2197–2202, 2003.

[16] D. L. Donoho, "For most large underdetermined systems of linear equations the minimal ℓ_1-norm solution is also the sparsest solution," *Communications on pure and applied mathematics*, vol. 59, no. 6, pp. 797–829, 2006.

[17] E. J. Candes and T. Tao, "Near-optimal signal recovery from random projections: Universal encoding strategies?" *IEEE Transactions on Information Theory*, vol. 52, no. 12, pp. 5406–5425, 2006.

[18] D. Donoho and J. Tanner, "Counting faces of randomly projected polytopes when the projection radically lowers dimension," *Journal of the American Mathematical Society*, vol. 22, no. 1, pp. 1–53, 2009.

[19] M. Lustig, D. L. Donoho, J. M. Santos, and J. M. Pauly, "Compressed sensing mri," *IEEE Signal Processing Magazine*, vol. 25, no. 2, pp. 72–82, 2008.

[20] M. F. Duarte, M. A. Davenport, D. Takhar, J. N. Laska, T. Sun, K. E. Kelly, and R. G. Baraniuk, "Single-pixel imaging via compressive sampling," *IEEE Signal Processing Magazine*, vol. 25, no. 2, p. 83, 2008.

[21] R. G. Baraniuk, V. Cevher, M. F. Duarte, and C. Hegde, "Model-based compressive sensing," *IEEE Transactions on Information Theory*, vol. 56, no. 4, pp. 1982–2001, 2010.

[22] D. Baron, S. Sarvotham, and R. G. Baraniuk, "Bayesian compressive sensing via belief propagation," *IEEE Transactions on Signal Processing*, vol. 58, no. 1, pp. 269–280, 2010.

[23] V. Cevher, P. Indyk, L. Carin, and R. G. Baraniuk, "Sparse signal recovery and acquisition with graphical models," *IEEE Signal Processing Magazine*, vol. 27, no. 6, pp. 92–103, 2010.

[24] A. K. Fletcher, S. Rangan, and V. K. Goyal, "On the rate-distortion performance of compressed sensing," in *2007 IEEE International Conference on Acoustics, Speech and Signal Processing (ICASSP)*, vol. 3. IEEE, 2007, pp. III–885.

[25] V. K. Goyal, A. K. Fletcher, and S. Rangan, "Compressive sampling and lossy compression," *IEEE Signal Processing Magazine*, vol. 25, no. 2, pp. 48–56, 2008.

[26] Y. Rachlin and D. Baron, "The secrecy of compressed sensing measurements," in *2008 Forty Sixth Annual Allerton Conference on Communication, Control, and Computing.* IEEE, 2008, pp. 813–817.

[27] J. Ranieri, R. Rovatti, and G. Setti, "Compressive sensing of localized signals: Application to analog-to-information conversion," in *2010 IEEE International Symposium on Circuits and Systems (ISCAS).* IEEE, 2010, pp. 3513–3516.

[28] M. Mangia, R. Rovatti, and G. Setti, "Analog-to-information conversion of sparse and non-white signals: Statistical design of sensing waveforms," in *2011 IEEE International Symposium on Circuits and Systems (ISCAS).* IEEE, 2011, pp. 2129–2132.

[29] ——, "Rakeness in the design of analog-to-information conversion of sparse and localized signals," *IEEE Transactions on Circuits and Systems I: Regular Papers*, vol. 59, no. 5, pp. 1001–1014, 2012.

[30] J. Romberg, "Compressive sensing by random convolution," *SIAM Journal on Imaging Sciences*, vol. 2, no. 4, pp. 1098–1128, 2009.

[31] H. Rauhut, J. Romberg, and J. A. Tropp, "Restricted isometries for partial random circulant matrices," *Applied and Computational Harmonic Analysis*, vol. 32, no. 2, pp. 242–254, 2012.

[32] T. Bjorklund and E. Magli, "A parallel compressive imaging architecture for one-shot acquisition," in *2013 IEEE Picture Coding Symposium (PCS)*, Dec 2013, pp. 65–68.

[33] R. M. Gray and D. L. Neuhoff, "Quantization," *IEEE Transactions on Information Theory*, vol. 44, no. 6, pp. 2325–2383, 1998.

[34] R. M. Gray, "Vector quantization," *IEEE Acoustics, Speech and Signal Processing Magazine*, vol. 1, no. 2, pp. 4–29, 1984.

[35] W. Rudin, *Real and complex analysis.* McGraw-Hill, 1987.

[36] C. E. Shannon, "Communication in the presence of noise," *Proceedings of the IRE*, vol. 37, no. 1, pp. 10–21, 1949.

[37] D. Slepian, "On bandwidth," *Proceedings of the IEEE*, vol. 64, no. 3, pp. 292–300, 1976.

[38] E. J. Candès and M. B. Wakin, "An introduction to compressive sampling," *IEEE Signal Processing Magazine*, vol. 25, no. 2, pp. 21–30, 2008.

[39] A. N. Kolmogorov, "Three approaches to the definition of the quantity of information," *Problemy Peredachi Informatsii*, vol. 1, no. 1, pp. 3–11, 1965.

[40] A. M. Turing, "On computable numbers, with an application to the entscheidungsproblem," *Proceedings of the London Mathematical Society*, vol. 2, no. 1, pp. 230–265, 1937.

[41] C. E. Shannon, "A mathematical theory of communication," *Bell System Technical Journal*, vol. 27, no. 4, pp. 623–656, 1948.

[42] P. D. Grünwald, *The minimum description length principle*. MIT Press, 2007.

[43] R. G. Baraniuk, V. Cevher, and M. B. Wakin, "Low-dimensional models for dimensionality reduction and signal recovery: A geometric perspective," *Proceedings of the IEEE*, vol. 98, no. 6, pp. 959–971, 2010.

[44] F. Bach, R. Jenatton, J. Mairal, and G. Obozinski, "Structured sparsity through convex optimization," *Statistical Science*, vol. 27, no. 4, pp. 450–468, 2012.

[45] S. Nam, M. E. Davies, M. Elad, and R. Gribonval, "The cosparse analysis model and algorithms," *Applied and Computational Harmonic Analysis*, vol. 34, no. 1, pp. 30–56, 2013.

[46] S. G. Mallat and Z. Zhang, "Matching pursuits with time-frequency dictionaries," *IEEE Transactions on Signal Processing*, vol. 41, no. 12, pp. 3397–3415, 1993.

[47] B. K. Natarajan, "Sparse approximate solutions to linear systems," *SIAM Journal on Computing*, vol. 24, no. 2, pp. 227–234, 1995.

[48] M. Elad, P. Milanfar, and R. Rubinstein, "Analysis versus synthesis in signal priors," *Inverse problems*, vol. 23, no. 3, p. 947, 2007.

[49] I. W. Selesnick and M. A. Figueiredo, "Signal restoration with overcomplete wavelet transforms: comparison of analysis and synthesis priors," in *SPIE Optical Engineering+ Applications*. International Society for Optics and Photonics, 2009, pp. 74 460D–74 460D.

[50] E. J. Candes, Y. C. Eldar, D. Needell, and P. Randall, "Compressed sensing with coherent and redundant dictionaries," *Applied and Computational Harmonic Analysis*, vol. 31, no. 1, pp. 59–73, 2011.

[51] A. Cohen, W. Dahmen, and R. DeVore, "Compressed sensing and best k-term approximation," *Journal of the American Mathematical Society*, vol. 22, no. 1, pp. 211–231, 2009.

[52] S. Mallat, *A wavelet tour of signal processing: the sparse way*. Access Online via Elsevier, 2008.

[53] L. I. Rudin, S. Osher, and E. Fatemi, "Nonlinear total variation based noise removal algorithms," *Physica D: Nonlinear Phenomena*, vol. 60, no. 1, pp. 259–268, 1992.

[54] M. Duarte, S. Sarvotham, D. Baron, M. Wakin, and R. Baraniuk, "Distributed compressed sensing of jointly sparse signals," in *2005 Thirty Ninth Asilomar Conference on Signals, Systems and Computers (ASILOMAR)*. IEEE, 2005, pp. 1537–1541.

[55] M. Wakin, M. Duarte, S. Sarvotham, D. Baron, and R. Baraniuk, "Recovery of jointly sparse signals from few random projections," in *Advances in Neural Information Processing Systems*, 2005, pp. 1433–1440.

[56] K. Karhunen, "Über lineare Methoden in der Wahrscheinlichkeitsrechnung," *Annales Academiae Scientiarum Fennicae Ser. A I Mathematica-Physica*, vol. 1947, no. 37, p. 79, 1947.

[57] G. Golub and W. Kahan, "Calculating the singular values and pseudo-inverse of a matrix," *Journal of the Society for Industrial & Applied Mathematics, Series B: Numerical Analysis*, vol. 2, no. 2, pp. 205–224, 1965.

[58] G. W. Stewart, "On the early history of the singular value decomposition," *SIAM Review*, vol. 35, no. 4, pp. 551–566, 1993.

[59] R. A. Horn and C. R. Johnson, *Matrix analysis*. Cambridge University Press, 2012.

[60] C. Eckart and G. Young, "The approximation of one matrix by another of lower rank," *Psychometrika*, vol. 1, no. 3, pp. 211–218, 1936.

[61] T. Skauli and J. Farrell, "A collection of hyperspectral images for imaging systems research," in *IS&T/SPIE Electronic Imaging*. International Society for Optics and Photonics, 2013, pp. 86 600C–86 600C.

[62] M. Vetterli and C. Herley, "Wavelets and filter banks: Theory and design," *IEEE Transactions on Signal Processing*, vol. 40, no. 9, pp. 2207–2232, 1992.

[63] T. H. Kim, H. J. Kong, T. H. Kim, and J. S. Shin, "Design and fabrication of a 900–1700nm hyper-spectral imaging spectrometer," *Optics Communications*, vol. 283, no. 3, pp. 355–361, 2010.

[64] E. J. Candès, "The restricted isometry property and its implications for compressed sensing," *Comptes Rendus Mathematique*, vol. 346, no. 9, pp. 589–592, 2008.

[65] R. Baraniuk, M. Davenport, R. DeVore, and M. Wakin, "A simple proof of the restricted isometry property for random matrices," *Constructive Approximation*, vol. 28, no. 3, pp. 253–263, 2008.

[66] A. Tillmann and M. Pfetsch, "The computational complexity of the restricted isometry property, the nullspace property, and related concepts in compressed sensing," *IEEE Transactions on Information Theory*, vol. 60, no. 2, pp. 1248–1259, Feb 2014.

[67] D. Achlioptas, "Database-friendly random projections: Johnson-lindenstrauss with binary coins," *Journal of Computer and System Sciences*, vol. 66, no. 4, pp. 671–687, 2003.

[68] S. Mendelson, A. Pajor, and N. Tomczak-Jaegermann, "Uniform uncertainty principle for Bernoulli and subgaussian ensembles," *Constructive Approximation*, vol. 28, no. 3, pp. 277–289, 2008.

[69] R. A. DeVore, "Deterministic constructions of compressed sensing matrices," *Journal of Complexity*, vol. 23, no. 4, pp. 918–925, 2007.

[70] R. Calderbank, S. Howard, and S. Jafarpour, "Construction of a large class of deterministic sensing matrices that satisfy a statistical isometry property," *IEEE Journal of Selected Topics in Signal Processing*, vol. 4, no. 2, pp. 358–374, 2010.

[71] J. Bourgain, S. Dilworth, K. Ford, S. Konyagin, and D. Kutzarova, "Explicit Constructions of RIP Matrices and Related Problems," *Duke Mathematical Journal*, vol. 159, no. 1, p. 145, 2011.

[72] A. S. Bandeira, M. Fickus, D. G. Mixon, and P. Wong, "The road to deterministic matrices with the restricted isometry property," *Journal of Fourier Analysis and Applications*, vol. 19, no. 6, pp. 1123–1149, 2013.

[73] J. D. Blanchard, C. Cartis, and J. Tanner, "Compressed sensing: How sharp is the restricted isometry property?" *SIAM Review*, vol. 53, no. 1, pp. 105–125, 2011.

[74] M. Mishali and Y. Eldar, "Expected RIP: Conditioning of The modulated wideband converter," in *2009 IEEE Information Theory Workshop (ITW)*, Oct 2009, pp. 343–347.

[75] L. Welch, "Lower bounds on the maximum cross correlation of signals (corresp.)," *IEEE Transactions on Information Theory*, vol. 20, no. 3, pp. 397–399, 1974.

[76] J. Massey, "On Welch's Bound for the Correlation of a Sequence Set," in *1991 IEEE International Symposium on Information Theory (ISIT)*, Jun 1991, pp. 385–385.

[77] S. A. Gershgorin, "Uber die abgrenzung der eigenwerte einer matrix," *Izvestiya Rossiiskoi Akademii Nauk, Seriya Matematicheskaya*, no. 6, pp. 749–754, 1931.

[78] J. A. Tropp, "Greed is good: Algorithmic results for sparse approximation," *IEEE Transactions on Information Theory*, vol. 50, no. 10, pp. 2231–2242, 2004.

[79] D. Donoho and J. Tanner, "Observed universality of phase transitions in high-dimensional geometry, with implications for modern data analysis and signal processing," *Philosophical Transactions of the Royal Society of London A: Mathematical, Physical and Engineering Sciences*, vol. 367, no. 1906, pp. 4273–4293, 2009.

[80] H. Rauhut, K. Schnass, and P. Vandergheynst, "Compressed sensing and redundant dictionaries," *IEEE Transactions on Information Theory*, vol. 54, no. 5, pp. 2210–2219, 2008.

[81] H. Rauhut, "Compressive sensing and structured random matrices," in *Theoretical Foundations and Numerical Methods for Sparse Recovery*. De Gruyter, 2010, vol. 9, pp. 1–92.

[82] R. Vershynin, "Introduction to the non-asymptotic analysis of random matrices," in *Compressed Sensing: Theory and Applications*. Cambridge University Press, 2012, pp. 210–268.

[83] S. Geman, "A limit theorem for the norm of random matrices," *The Annals of Probability*, vol. 8, no. 2, pp. 252–261, 1980.

[84] J. W. Silverstein, "The smallest eigenvalue of a large dimensional Wishart matrix," *The Annals of Probability*, vol. 13, no. 4, pp. 1364–1368, 1985.

[85] Z. Bai and Y. Yin, "Limit of the smallest eigenvalue of a large dimensional sample covariance matrix," *The Annals of Probability*, pp. 1275–1294, 1993.

[86] V. A. Marchenko and L. A. Pastur, "Distribution of eigenvalues for some sets of random matrices," *Sbornik: Mathematics*, vol. 1, no. 4, pp. 457–483, 1967.

[87] N. Ailon and B. Chazelle, "Approximate nearest neighbors and the fast Johnson-Lindenstrauss transform," in *Proceedings of the thirty-eighth annual ACM Symposium on Theory of Computing*. ACM, 2006, pp. 557–563.

[88] ——, "The fast Johnson-Lindenstrauss transform and approximate nearest neighbors," *SIAM Journal on Computing*, vol. 39, no. 1, pp. 302–322, 2009.

[89] G. Puy, P. Vandergheynst, R. Gribonval, and Y. Wiaux, "Universal and efficient compressed sensing by spread spectrum and application to realistic fourier imaging techniques," *EURASIP Journal on Advances in Signal Processing*, vol. 2012, no. 1, pp. 1–13, 2012.

[90] D. L. Donoho and J. Tanner, "Precise undersampling theorems," *Proceedings of the IEEE*, vol. 98, no. 6, pp. 913–924, 2010.

[91] M. A. Herman and T. Strohmer, "General deviants: An analysis of perturbations in compressed sensing," *IEEE Journal of Selected Topics in Signal Processing*, vol. 4, no. 2, pp. 342–349, 2010.

[92] M. Grant and S. Boyd, "CVX: Matlab software for disciplined convex programming, version 2.1," http://cvxr.com/cvx, Mar. 2015.

[93] Gurobi Optimization, Inc., "Gurobi optimizer reference manual," 2015. [Online]. Available: http://www.gurobi.com

[94] ILOG, Inc., "ILOG CPLEX: High-performance software for mathematical programming and optimization," 2015, http://www.ilog.com/products/cplex/.

[95] S. P. Boyd and L. Vandenberghe, *Convex optimization*. Cambridge University Press, 2004.

[96] P. L. Combettes and V. R. Wajs, "Signal recovery by proximal forward-backward splitting," *Multiscale Modeling & Simulation*, vol. 4, no. 4, pp. 1168–1200, 2005.

[97] N. Parikh and S. Boyd, "Proximal algorithms," *Foundations and Trends in Optimization*, pp. 1–96, 2013.

[98] E. van den Berg and M. P. Friedlander, "Probing the pareto frontier for basis pursuit solutions," *SIAM Journal on Scientific Computing*, vol. 31, no. 2, pp. 890–912, 2008. [Online]. Available: http://link.aip.org/link/?SCE/31/890

[99] D. Needell and J. A. Tropp, "Cosamp: Iterative signal recovery from incomplete and inaccurate samples," *Applied and Computational Harmonic Analysis*, vol. 26, no. 3, pp. 301–321, 2009.

[100] T. Blumensath and M. E. Davies, "Iterative thresholding for sparse approximations," *Journal of Fourier Analysis and Applications*, vol. 14, no. 5-6, pp. 629–654, 2008.

[101] A. Maleki and D. L. Donoho, "Optimally tuned iterative reconstruction algorithms for compressed sensing," *IEEE Journal of Selected Topics in Signal Processing*, vol. 4, no. 2, pp. 330–341, 2010.

[102] J. A. Tropp and S. J. Wright, "Computational methods for sparse solution of linear inverse problems," *Proceedings of the IEEE*, vol. 98, no. 6, pp. 948–958, 2010.

[103] S. Ji, Y. Xue, and L. Carin, "Bayesian compressive sensing," *IEEE Transactions on Signal Processing*, vol. 56, no. 6, pp. 2346–2356, 2008.

[104] D. L. Donoho, A. Maleki, and A. Montanari, "Message-passing algorithms for compressed sensing," *Proceedings of the National Academy of Sciences*, vol. 106, no. 45, pp. 18 914–18 919, 2009.

[105] S. Rangan, "Generalized approximate message passing for estimation with random linear mixing," in *2011 IEEE International Symposium on Information Theory (ISIT)*. IEEE, 2011, pp. 2168–2172.

[106] R. Gribonval and M. Nielsen, "Sparse representations in unions of bases," *IEEE Transactions on Information Theory*, vol. 49, no. 12, pp. 3320–3325, 2003.

[107] Y. Wang and W. Yin, "Sparse signal reconstruction via iterative support detection," *SIAM Journal on Imaging Sciences*, vol. 3, no. 3, pp. 462–491, 2010.

[108] J. A. Tropp and A. C. Gilbert, "Signal recovery from random measurements via orthogonal matching pursuit," *IEEE Transactions on Information Theory*, vol. 53, no. 12, pp. 4655–4666, 2007.

[109] R. Tibshirani, "Regression shrinkage and selection via the LASSO," *Journal of the Royal Statistical Society, Series B (Methodological)*, pp. 267–288, 1996.

[110] P. C. Hansen, "Analysis of discrete ill-posed problems by means of the l-curve," *SIAM Review*, vol. 34, no. 4, pp. 561–580, 1992.

[111] E. van den Berg and M. P. Friedlander, "SPGL1: A solver for large-scale sparse reconstruction," June 2007, http://www.cs.ubc.ca/labs/scl/spgl1.

[112] L. Vandenberghe and S. Boyd, "Semidefinite programming," *SIAM Review*, vol. 38, no. 1, pp. 49–95, 1996.

[113] P. L. Combettes and J.-C. Pesquet, "A douglas–rachford splitting approach to nonsmooth convex variational signal recovery," *IEEE Journal of Selected Topics in Signal Processing*, vol. 1, no. 4, pp. 564–574, 2007.

[114] M. J. Fadili and J. L. Starck, "Monotone operator splitting for optimization problems in sparse recovery," in *2009 IEEE International Conference on Image Processing (ICIP)*. IEEE, 2009, pp. 1461–1464.

[115] Z. Zhang and B. D. Rao, "Sparse signal recovery with temporally correlated source vectors using sparse bayesian learning," *IEEE Journal of Selected Topics in Signal Processing*, vol. 5, no. 5, pp. 912–926, 2011.

[116] M. Mangia, J. Haboba, R. Rovatti, and G. Setti, "Rakeness-based approach to compressed sensing of ecgs," in *2011 IEEE Biomedical Circuits and Systems Conference (BioCAS)*. IEEE, 2011, pp. 424–427.

[117] P. Billingsley, *Probability and measure*. John Wiley & Sons, 2008.

[118] R. Gribonval, V. Cevher, and M. E. Davies, "Compressible distributions for high-dimensional statistics," *IEEE Transactions on Information Theory*, vol. 58, no. 8, pp. 5016–5034, 2012.

[119] Y. Gordon, "Elliptically contoured distributions," *Probability theory and related fields*, vol. 76, no. 4, pp. 429–438, 1987.

[120] A. L. Goldberger, L. A. N. Amaral, L. Glass, J. M. Hausdorff, P. C. Ivanov, R. G. Mark, J. E. Mietus, G. B. Moody, C.-K. Peng, and H. E. Stanley, "PhysioBank, PhysioToolkit, and PhysioNet: Components of a new research resource for complex physiologic signals," *Circulation*, vol. 101, no. 23, pp. 215–220, Jun. 2000.

[121] S. Cassidy and J. Harrington, "The emu speech database system," 2000, available at http://emu.sourceforge.net/.

[122] A. J. Hoffman and H. W. Wielandt, "The variation of the spectrum of a normal matrix," *Duke Mathematical Journal*, vol. 20, no. 1, pp. 37–39, 1953.

[123] R. Rovatti, G. Mazzini, G. Setti, and S. Vitali, "Linear probability feedback processes," in *2008 IEEE International Symposium on Circuits and Systems (ISCAS)*. IEEE, 2008, pp. 548–551.

[124] R. Rovatti, G. Mazzini, and G. Setti, "Memory-m antipodal processes: spectral analysis and synthesis," *IEEE Transactions on Circuits and Systems I: Regular Papers*, vol. 56, no. 1, 2009.

[125] A. Caprara, F. Furini, A. Lodi, M. Mangia, R. Rovatti, and G. Setti, "Generation of antipodal random vectors with prescribed non-stationary 2-nd order statistics," *IEEE Transactions on Signal Processing*, vol. 62, no. 6, pp. 1603–1612, March 2014.

[126] J. H. Van Vleck and D. Middleton, "The spectrum of clipped noise," *Proceedings of the IEEE*, vol. 54, no. 1, pp. 2–19, 1966.

[127] G. Jacovitti, A. Neri, and G. Scarano, "Texture synthesis-by-analysis with hard-limited gaussian processes," *IEEE Transactions on Image Processing*, vol. 7, no. 11, pp. 1615–1621, 1998.

[128] F. Zhang, *Matrix theory: basic results and techniques*. Springer, 2011.

[129] I. F. Akyildiz, W. Su, Y. Sankarasubramaniam, and E. Cayirci, "A survey on sensor networks," *IEEE Communications Magazine*, vol. 40, no. 8, pp. 102–114, 2002.

[130] T. M. Cover and J. A. Thomas, *Elements of information theory*. John Wiley & Sons, 2012.

[131] W. A. Pearlman and A. Said, *Digital Signal Compression: Principles and Practice*. Cambridge University Press, 2011.

[132] Z. Lu, D. Y. Kim, and W. A. Pearlman, "Wavelet compression of ecg signals by the set partitioning in hierarchical trees algorithm," *IEEE Transactions on Biomedical Engineering*, vol. 47, no. 7, pp. 849–856, 2000.

[133] K. Kotteri, S. Barua, A. Bell, and J. Carletta, "A comparison of hardware implementations of the biorthogonal 9/7 dwt: convolution versus lifting," *IEEE Transactions on Circuits and Systems II: Express Briefs*, vol. 52, no. 5, pp. 256–260, 2005.

[134] A. R. Calderbank, I. Daubechies, W. Sweldens, and B.-L. Yeo, "Wavelet Transforms That Map Integers to Integers," *Applied and Computational Harmonic Analysis*, vol. 5, no. 3, pp. 332–369, Jul. 1998.

[135] J. Max, "Quantizing for minimum distortion," *IRE Transactions on Information Theory*, vol. 6, no. 1, pp. 7–12, March 1960.

[136] J. Z. Sun and V. K. Goyal, "Optimal quantization of random measurements in compressed sensing," in *2009 IEEE International Symposium on Information Theory (ISIT)*. IEEE, 2009, pp. 6–10.

[137] L. Jacques, D. Hammond, and M. Fadili, "Stabilizing Nonuniformly Quantized Compressed Sensing With Scalar Companders," *IEEE Transactions on Information Theory*, vol. 59, no. 12, pp. 7969–7984, Dec. 2013.

[138] V. K. Goyal, "Theoretical foundations of transform coding," *IEEE Signal Processing Magazine*, vol. 18, no. 5, pp. 9–21, 2001.

[139] J. E. Fowler, "The redundant discrete wavelet transform and additive noise," *IEEE Signal Processing Letters*, vol. 12, no. 9, pp. 629–632, 2005.

[140] P. E. McSharry, G. D. Clifford, L. Tarassenko, and L. A. Smith, "A dynamical model for generating synthetic electrocardiogram signals," *IEEE Transactions on Biomedical Engineering*, vol. 50, no. 3, pp. 289–294, 2003.

[141] N. Perraudin, D. Shuman, G. Puy, and P. Vandergheynst, "UNLocBoX: A matlab convex optimization toolbox using proximal splitting methods," *arXiv preprint arXiv:1402.0779*, 2014.

[142] L. Jacques, D. K. Hammond, and J. M. Fadili, "Dequantizing compressed sensing: When oversampling and non-gaussian constraints combine," *IEEE Transactions on Information Theory*, vol. 57, no. 1, pp. 559–571, 2011.

[143] E. T. Jaynes, "On the rationale of maximum-entropy methods," *Proceedings of the IEEE*, vol. 70, no. 9, pp. 939–952, 1982.

[144] B. J. Fino and V. Algazi, "Unified matrix treatment of the fast walsh-hadamard transform," *IEEE Transactions on Computers*, vol. C-25, no. 11, pp. 1142–1146, 1976.

[145] J. Portilla, V. Strela, M. J. Wainwright, and E. P. Simoncelli, "Image denoising using scale mixtures of gaussians in the wavelet domain," *IEEE Transactions on Image Processing*, vol. 12, no. 11, pp. 1338–1351, 2003.

[146] M. Huber, T. Bailey, H. Durrant-Whyte, and U. D. Hanebeck, "On entropy approximation for gaussian mixture random vectors," in *2008 IEEE International Conference on Multisensor Fusion and Integration for Intelligent Systems (MFI)*, 2008, pp. 181–188.

[147] P. P. Vaidyanathan, "The theory of linear prediction," *Synthesis Lectures on Signal Processing*, vol. 2, no. 1, pp. 1–184, 2007.

[148] C.-W. Ko, J. Lee, and M. Queyranne, "An exact algorithm for maximum entropy sampling," *Operations Research*, vol. 43, no. 4, pp. 684–691, 1995.

[149] M. Mitchell, *An Introduction to Genetic Algorithms (Complex Adaptive Systems)*. A Bradford Book, Feb. 1998.

[150] J. J. Hull, "A database for handwritten text recognition research," *IEEE Transactions on Pattern Analysis and Machine Intelligence*, vol. 16, no. 5, pp. 550–554, 1994.

[151] R. M. Willett, R. F. Marcia, and J. M. Nichols, "Compressed sensing for practical optical imaging systems: a tutorial," *Optical Engineering*, vol. 50, no. 7, pp. 072 601–072 601, 2011.

[152] M. Mangia, F. Pareschi, R. Rovatti, and G. Setti, "Leakage compensation in analog random modulation pre-integration architectures for biosignal acquisition," in *2014 IEEE Biomedical Circuits and Systems Conference (BioCAS)*. IEEE, Oct. 2014.

[153] C. Schulke, F. Caltagirone, F. Krzakala, and L. Zdeborova, "Blind calibration in compressed sensing using message passing algorithms," in *Advances in Neural Information Processing Systems*, 2013, pp. 566–574.

[154] C. Bilen, G. Puy, R. Gribonval, and L. Daudet, "Convex optimization approaches for blind sensor calibration using sparsity," *IEEE Transactions on Signal Processing*, vol. 62, no. 18, pp. 4847–4856, Sept. 2014.

[155] Z. Ben-Haim and Y. C. Eldar, "Performance bounds for sparse estimation with random noise," in *2009 IEEE Statistical Signal Processing Workshop (SSP)*. IEEE, 2009, pp. 225–228.

[156] ——, "The Cramér-Rao bound for estimating a sparse parameter vector," *IEEE Transactions on Signal Processing*, vol. 58, no. 6, pp. 3384–3389, 2010.

[157] P.-L. Loh and M. J. Wainwright, "High-dimensional regression with noisy and missing data: Provable guarantees with nonconvexity," *Annals of Statistics*, vol. 40, no. 3, p. 1637, 2012.

[158] ——, "Corrupted and missing predictors: Minimax bounds for high-dimensional linear regression," in *2012 IEEE International Symposium on Information Theory (ISIT)*. IEEE, 2012, pp. 2601–2605.

[159] H. Zhu, G. Leus, and G. B. Giannakis, "Sparsity-cognizant total least-squares for perturbed compressive sampling," *IEEE Transactions on Signal Processing*, vol. 59, no. 5, pp. 2002–2016, 2011.

[160] J. T. Parker, V. Cevher, and P. Schniter, "Compressive sensing under matrix uncertainties: An approximate message passing approach," in *2011 Forty Fifth Asilomar Conference on Signals, Systems and Computers (ASILOMAR)*. IEEE, 2011, pp. 804–808.

[161] Y. Wang, G. Attebury, and B. Ramamurthy, "A survey of security issues in wireless sensor networks," *IEEE Communications Surveys & Tutorials*, vol. 8, no. 2, pp. 2–23, 2006.

[162] J. Daemen and V. Rijmen, *The design of Rijndael: AES-the advanced encryption standard*. Springer, 2002.

[163] A. Orsdemir, H. O. Altun, G. Sharma, and M. F. Bocko, "On the security and robustness of encryption via compressed sensing," in *2008 IEEE Military Communications Conference (MILCOM)*. IEEE, 2008, pp. 1–7.

[164] L. C. Washington and W. Trappe, *Introduction to cryptography: with coding theory*. Prentice Hall – PTR, 2002.

[165] I. Drori, "Compressed video sensing," in *2008 BMVA Symposium on 3D Video-Analysis, Display, and Applications*, 2008.

[166] J. Massey, "Shift-register synthesis and bch decoding," *IEEE Transactions on Information Theory*, vol. 15, no. 1, pp. 122–127, 1969.

[167] M. Matsumoto, T. Nishimura, M. Saito, and M. Hagita, "Cryptographic mersenne twister and fubuki stream/block cipher," *Cryptographic ePrint Archive*, 2005.

[168] D. Eastlake and P. Jones, "US Secure Hash Algorithm 1 (SHA1)," Sep. 2001.

[169] R. Latała, "Some estimates of norms of random matrices," *Proceedings of the American Mathematical Society*, vol. 133, no. 5, pp. 1273–1282, 2005.

[170] J. Vila and P. Schniter, "Expectation-maximization Bernoulli-Gaussian approximate message passing," in *2011 Forty Fifth Asilomar Conference on Signals, Systems and Computers (ASILOMAR)*. IEEE, 2011, pp. 799–803.

[171] G. Pirker, M. Wohlmayr, S. Petrik, and F. Pernkopf, "A Pitch Tracking Corpus with Evaluation on Multipitch Tracking Scenario," in *Interspeech 2011*, Florence (Italy), Aug. 27-31, 2011, pp. 1509–1512.

[172] G. Shires and H. Wennborg, "Web Speech API Specification," Oct. 2012, http://dvcs.w3.org/hg/speech-api/raw-file/tip/speechapi.html.

[173] G. Hinton, L. Deng, D. Yu, G. E. Dahl, A.-r. Mohamed, N. Jaitly, A. Senior, V. Vanhoucke, P. Nguyen, and T. N. Sainath, "Deep neural networks for acoustic modeling in speech recognition: the shared views of four research groups," *IEEE Signal Processing Magazine*, vol. 29, no. 6, pp. 82–97, 2012.

[174] R. Jane, A. Blasi, J. García, and P. Laguna, "Evaluation of an automatic threshold based detector of waveform limits in holter ecg with the qt database," in *1997 Computers in Cardiology Conference (CINC)*. IEEE, 1997, pp. 295–298.

[175] R. Smith, "An Overview of the Tesseract OCR Engine," in *2007 Ninth International Conference on Document Analysis and Recognition (ICDAR)*, vol. 2, 2007, pp. 629–633.

[176] C. E. Shannon, "Communication theory of secrecy systems," *Bell System Technical Journal*, vol. 28, no. 4, pp. 656–715, 1949.

[177] T. Bianchi, V. Bioglio, and E. Magli, "On the security of random linear measurements," in *2014 IEEE International Conference on Acoustics, Speech and Signal Processing (ICASSP)*. IEEE, 2014, pp. 3992–3996.

[178] A. Wyner, "The wire-tap channel," *Bell System Technical Journal*, vol. 54, no. 8, pp. 1355–1387, Oct 1975.

[179] A. C. Berry, "The accuracy of the gaussian approximation to the sum of independent variates," *Transactions of the American Mathematical Society*, vol. 49, no. 1, pp. 122–136, 1941.

[180] B. Klartag and S. Sodin, "Variations on the berry–esseen theorem," *Theory of Probability & Its Applications*, vol. 56, no. 3, pp. 403–419, 2012.

[181] A. Kerckhoffs, "La cryptographie militaire," *Journal des sciences militaires*, vol. IX, pp. 5–38, Jan. 1883.

[182] S. Martello and P. Toth, *Knapsack problems: algorithms and computer implementations*. John Wiley & Sons, Inc., 1990.

[183] J. C. Lagarias and A. M. Odlyzko, "Solving low-density subset sum problems," *Journal of the ACM (JACM)*, vol. 32, no. 1, pp. 229–246, 1985.

[184] R. Merkle and M. Hellman, "Hiding information and signatures in trapdoor knapsacks," *IEEE Transactions on Information Theory*, vol. 24, no. 5, pp. 525–530, 1978.

[185] B. Chor and R. L. Rivest, "A knapsack-type public key cryptosystem based on arithmetic in finite fields," *IEEE Transactions on Information Theory*, vol. 34, no. 5, pp. 901–909, 1988.

[186] A. M. Odlyzko, "The rise and fall of knapsack cryptosystems," *Cryptology and Computational Number Theory*, vol. 42, pp. 75–88, 1990.

[187] T. Sasamoto, T. Toyoizumi, and H. Nishimori, "Statistical mechanics of an np-complete problem: subset sum," *Journal of Physics A: Mathematical and General*, vol. 34, no. 44, p. 9555, 2001.

[188] E. Ehrhart, "Sur un probleme de géométrie diophantienne linéaire. ii. systemes diophantiens linéaires. (french)," *Journal für die reine und angewandte Mathematik*, vol. 227, pp. 25–49, 1967.

[189] I. G. Macdonald, "Polynomials Associated with Finite Cell-Complexes," *Journal of the London Mathematical Society*, vol. 2, no. 1, pp. 181–192, 1971.

[190] P. Mouroulis and M. M. McKerns, "Pushbroom imaging spectrometer with high spectroscopic data fidelity: experimental demonstration," *Optical Engineering*, vol. 39, no. 3, pp. 808–816, 2000.

[191] J. Veverka, J. Bell, P. Thomas, A. Harch, S. Murchie, S. Hawkins, J. Warren, H. Darlington, K. Peacock, C. Chapman et al., "An overview of the near multispectral imager-near-infrared spectrometer investigation," *Journal of Geophysical Research: Planets (1991–2012)*, vol. 102, no. E10, pp. 23 709–23 727, 1997.

[192] W. L. Wolfe, *Introduction to imaging spectrometers*. SPIE Press, 1997, vol. 25.

[193] A. R. Harvey, J. E. Beale, A. H. Greenaway, T. J. Hanlon, and J. W. Williams, "Technology options for imaging spectrometry," in *International Symposium on Optical Science and Technology*. International Society for Optics and Photonics, 2000, pp. 13–24.

[194] B. Geelen, N. Tack, and A. Lambrechts, "A compact snapshot multispectral imager with a monolithically integrated per-pixel filter mosaic," in *SPIE MOEMS-MEMS*. International Society for Optics and Photonics, 2014, pp. 89 740L–89 740L.

[195] N. Tack, A. Lambrechts, P. Soussan, and L. Haspeslagh, "A compact, high-speed, and low-cost hyperspectral imager," in *SPIE OPTO*. International Society for Optics and Photonics, 2012, pp. 82 660Q–82 660Q.

[196] A. Lipson, S. G. Lipson, and H. Lipson, *Optical physics*. Cambridge University Press, 2010.

[197] A. Wagadarikar, R. John, R. Willett, and D. Brady, "Single disperser design for coded aperture snapshot spectral imaging," *Applied Optics*, vol. 47, no. 10, pp. B44–B51, 2008.

[198] G. Arce, D. Brady, L. Carin, H. Arguello, and D. Kittle, "Compressive coded aperture spectral imaging: An introduction," *IEEE Signal Processing Magazine*, vol. 31, no. 1, pp. 105–115, 2014.

[199] L. Jacques, P. Vandergheynst, A. Bibet, V. Majidzadeh, A. Schmid, and Y. Leblebici, "Cmos compressed imaging by random convolution," in *2009 IEEE International Conference on Acoustics, Speech and Signal Processing (ICASSP)*. IEEE, 2009, pp. 1113–1116.

[200] Y. Wu, I. O. Mirza, G. R. Arce, and D. W. Prather, "Development of a digital-micromirror-device-based multishot snapshot spectral imaging system," *Optics Letters*, vol. 36, no. 14, pp. 2692–2694, 2011.

[201] W. L. Chan, K. Charan, D. Takhar, K. F. Kelly, R. G. Baraniuk, and D. M. Mittleman, "A single-pixel terahertz imaging system based on compressed sensing," *Applied Physics Letters*, vol. 93, no. 12, p. 121105, 2008.

[202] R. F. Marcia, Z. T. Harmany, and R. M. Willett, "Compressive coded aperture imaging," in *IS&T/SPIE Electronic Imaging*. International Society for Optics and Photonics, 2009, pp. 72 460G–72 460G.

[203] J. Goodman, *Introduction to Fourier optics*. McGraw-hill, 2008.

[204] D. J. Brady, *Optical imaging and spectroscopy*. John Wiley & Sons, 2009.

[205] E. E. Fenimore and T. M. Cannon, "Coded aperture imaging with uniformly redundant arrays," *Applied Optics*, vol. 17, no. 3, pp. 337–347, 1978.

[206] A. K. Jain, *Fundamentals of digital image processing*. Prentice-Hall Englewood Cliffs, 1989, vol. 3.

[207] J. M. Bioucas-Dias and M. A. Figueiredo, "A new twist: two-step iterative shrinkage/thresholding algorithms for image restoration," *IEEE Transactions on Image Processing*, vol. 16, no. 12, pp. 2992–3004, 2007.

[208] M. Golbabaee and P. Vandergheynst, "Hyperspectral image compressed sensing via low-rank and joint-sparse matrix recovery," in *2012 IEEE International Conference on Acoustics, Speech and Signal Processing (ICASSP)*. IEEE, 2012, pp. 2741–2744.

[209] A. Chambolle and T. Pock, "A first-order primal-dual algorithm for convex problems with applications to imaging," *Journal of Mathematical Imaging and Vision*, vol. 40, no. 1, pp. 120–145, 2011.

[210] H. Nagahara, C. Zhou, T. Watanabe, H. Ishiguro, and S. K. Nayar, "Programmable aperture camera using lcos," in *2010 European Conference on Computer Vision (ECCV)*. Springer, 2010, pp. 337–350.

[211] G. R. Fowles, *Introduction to modern optics*. Courier Dover Publications, 2012.

[212] A. Liutkus, D. Martina, S. Popoff, G. Chardon, O. Katz, G. Lerosey, S. Gigan, L. Daudet, and I. Carron, "Imaging with nature: Compressive imaging using a multiply scattering medium," *Scientific Reports*, vol. 4, p. 5552, 2014.